THE KALMAN FILTER IN FINANCE

Advanced Studies in Theoretical and Applied Econometrics

Volume 32

The Kalman Filter
in Finance

by

Curt Wells
University of Lund,
Sweden

KLUWER ACADEMIC PUBLISHERS
DORDRECHT / BOSTON / LONDON

A C.I.P. Catalogue record for this book is available from the Library of Congress.

ISBN 0-7923-3771-9

Published by Kluwer Academic Publishers,
P.O. Box 17, 3300 AA Dordrecht, The Netherlands.

Kluwer Academic Publishers incorporates
the publishing programmes of
D. Reidel, Martinus Nijhoff, Dr W. Junk and MTP Press.

Sold and distributed in the U.S.A. and Canada
by Kluwer Academic Publishers,
101 Philip Drive, Norwell, MA 02061, U.S.A.

In all other countries, sold and distributed
by Kluwer Academic Publishers Group,
P.O. Box 322, 3300 AH Dordrecht, The Netherlands.

Printed on acid-free paper

Printed in the Netherlands

To the women in my life:
Ann-Charlotte
Ann-Catrin
Lisbeth

Contents

List of Figures

List of Tables

Preface

Work on the project that was to result in this book began at the end of the 1980's. We have a club for mutual admiration consisting of those of us in Sweden who work with econometric problems. We meet once a year in Lund or at one of the other four main universities of the kingdom. At one of the meetings, in the spring of 1989, I discussed the problem of estimating time–varying coefficients with one of the other participants. He said, "use the Kalman filter and program it in GAUSS. I'll even send you a copy of a *GAUSS Newsletter* on the subject." That conversation has resulted in the work reported here. In a way, this book is the logical conclusion to my dissertation which considered *linear quadratic optimization* except that in this book, parameter estimation rather than economic policy is optimized.

The book may be divided into three main parts. The first of these three are introductions. Chapter 1 contains a derivation and a historical description of *beta* in the literature and may be viewed as an introduction to this coefficient for the reader that is not well versed in Finance. The second chapter of an introductory nature is Chapter 4 which presents the Kalman filter. Here the emphasis is on understanding the filter rather than mathematics. The second part of the book considers tests for non constant regression coefficients and consists of Chapters 2 and 3. The first of these presents well known tests and applies them to data from the Stockholm Exchange. The second of the two chapters discusses a rather unfamiliar way of examining the data for parameter stability. The final three chapters make up the last part of the book. Here the issue is estimating models with time–varying parameters. Chapter 5 presents the main results; Chapter 6 takes a second look at them; Chapter 7 shows how these results could be applied to other data sets.

This book is targeted at graduate students in econometrics as well as researchers who wish to learn the practical aspects of estimating with the Kalman filter. There are indeed many fields where models with time–varying parameters should be relevant; some examples conclude the book but many more could be found. We consider only single equation models. Multidimensional systems such a macro models or models of exchange rates are instances where the time–varying parameters also seem applicable.

I expect that there are many who would like to use these methods but are balked by the initial cost of programming. I am by no means a computer expert; my code is straight forward, but it works and I am sure that others could find it useful. To this extent, I have made available the programs in the appendix on our *anonymous ftp*. The programs are in GAUSS and with the exception of the routine *maxlik* are self contained. Recently the people at Aptech have released a version of *maxlik* that does *constrained* maximum likelihood. The programs here are not designed for that application. To help get readers started, I have also included the data as GAUSS data sets – or in some cases, as ASCII files – on the *ftp*. The details are to be found in Appendix B on page 151.

A project that spans a number of years is always improved by the suggestions of others. I've had the opportunity to present my ideas at the 1991 meetings of the European Finance Association in Rotterdam, for the Graduate seminars at the Stockholm School of Economics and here in Lund. Comments received from the participants have been most helpful. On a more mundane level, I should thank Tore Browaldh's Foundation for Research in the Social Sciences for the support given that has, in part, made research on the book possible. Erik Eklund at *Findata* in Stockholm has helped by supplying the data on Swedish stocks. Prof. Timo Teräsvirta took the time to comment on Chapter 2. My colleague David Edgerton has also provided comments on many of the statistical procedures used in that same chapter. Prof. Wolfgang Schneider provided the GAUSS code for the *FLS* estimations in Chapter 3. However, I have modified his programs using the original *FORTRAN* routines provided by Prof. Tesfatsion. I should also note lectures given by Prof. Schneider at the Stockholm School of Economics in the Spring of 1990 have been a source of inspiration for my work. The manuscript has been redone from one of the commercial desktop publishing programs into LaTeX. But even here I have had help: David Carlisle has been kind enough to answer my silly questions about the inner workings of LaTeX. I have even thrown in the odd semicolon for Mari.

A final thanks must of course go to my wife and my two little daughters who have had to put up with both my coming home late at night and my work at the computer in the summers. They learned that it was useless to approach me when I sat facing my monitor; they may not always have liked it, but they accepted it as how things were. I do suspect, however, that they are thanking the powers that be for the completion of this book.

Curt Wells
Lund, Summer 1995

Chapter 1

Introduction

1.1. The prelude

As a beginning student of economics, I learned that economic coefficients were constants. A good example is provided by a passage in a introductory textbook on macro economics where Prof. Ackley "demonstrated" how he could use a "sex multiplier" to calculate changes in the enrollment in his economics class. (See Ackley (1961), pp. 309-312.) He would simply multiply the change in number of female students by his multiplier to get the change in total enrollment. As there were fewer women than men, this model saved him time. Of course, different years had different multipliers, but in the end it always worked. The point that Ackley was making was the difference between meaningful forecasts using *ex ante* models and tautologies derived from *ex post* calculations. "If we merely had some reason ... to assert the probable *constancy* or *stability* of the [sex multiplier] ... then our theory would be ... potentially useful", writes Ackley. The message is that theory should provide stable parameters that our analysis of the economy should be based on.[1]

Econometric models on the whole assume constant parameters. Keynes, commenting on a draft of Tinbergen's Business cycles, reportedly wrote:

> "The coefficients arrived at are apparently assumed to be constant for 10 years or for a longer period. Yet, surely we know that they are not constant. There is no reason at all why they should not be different every year." (Moggridge (1973), p. 286)

[1]This is perhaps unfair criticism; in 1961, Kalman's work was still new and under development and the methodology needed for the estimation of models with time–varying parameters had yet to be perfected. Still, as a student, I learned that the concept of varying parameters was close to heresy. Prof. Ackley's comments must have, however, impressed me as I recalled the passage some 25 years after reading it. Or perhaps it was the catchy name of his multiplier that stuck in my memory.

Keynes was of course referring to annual models, but the point is clear: why should coefficients be constant? At the very least, the problems associated with aggregation from the micro unit to the macro unit would suggest that varying coefficient models are to be preferred. (see Swamy *et al.*, 1988)

This book is concerned with the estimation of time–varying parameters. Given today's state of the art in personal computers and software, the methods outlined here are available to all, not just an esoteric few. The main model to be studied is the *Capital Asset Pricing Model* – the *CAPM* of basic finance. This has been chosen not only because of the abundance of financial data, but also because the profession's attitude towards the *beta* coefficient, which represents market risk, has evolved from an initial hypothesis of stability to one of non constancy.[2] The remainder of this chapter, after a brief introduction to the *CAPM*, will try to trace this changing attitude.

The rest of the book will first test for constant coefficients and then estimate the models with time–varying parameters. Chapter 2 presents and performs a number of statistical tests on the data; Chapter 3 adds another way of examining the question of parameter stability. Chapter 4 provides a rather informal introduction to the Kalman filter and Chapter 5 applies the filter to the problem of estimating the *CAPM*. The book concludes with two chapters: one that reexamines some of the estimates presented and one that shows how the methods of Chapters 2–5 may be applied to a number of other models.

1.2. The Market Line

I will begin by presenting a rather heuristic derivation of the equation that will occupy our attention through much of this book.[3] Let us assume that an individual plans to invest part of his wealth in a risk free asset and the remaining part in a risky asset. Let r_f be the expected returns to the risk free asset and r_m be those to the risky one. If he invests the fraction x of his wealth in the risky asset, then the expected return to his investment will be

$$r_x = (1 - x)r_f + xr_m$$

The risk of holding an asset is measured by the standard deviation in the returns to that asset, σ_m. Thus the variance of the risk free asset must be

---

[2]Macro economic modeling has taken other routes: rather than estimating time–varying coefficients, *ARCH* and *GARCH* methods model time–varying error variances; *VAR* models take a time–series approach to modeling where results rather than parameters are important. Even non–linear models such as the *STAR* model are formulated and estimated. This book will concentrate on time–varying parameters in linear models.

[3]My presentation follows Varian (1993), pp. 230–239. Consult this excellent text for greater detail.

zero while the that of the risky asset is assumed to be σ_m^2. As the returns to the risky asset are assumed independent of those of the risky one, the variance of his investment will be:

$$\text{var}(r_x) \equiv \sigma_x^2 = (1-x)^2 \cdot 0 + x^2\sigma_m^2 = x^2\sigma_m^2$$

Finally, assume that the investor's utility is a function of two arguments, the return to his investment and the risk inherent in that investment:

$$u\left(r_x, \sigma_x\right) = u\left((1-x)\,r_f + xr_m, x\sigma_x\right) \tag{1.1}$$

In equation (1.1), utility has been expressed in terms of the expected returns to the two assets, the variance in the risky asset and the proportion of his wealth that is invested in this risky asset. Note that while returns are a "good", risk is a 'bad" in that utility increases as risk *decreases*. We assume that he chooses x so that his utility is maximized. The first order conditions are

$$\frac{\partial u}{\partial r_x}\left(r_m - r_f\right) + \frac{\partial u}{\partial \sigma_x}\sigma_m = 0$$

Rearranging gives us a familiar relationship:

$$\frac{\partial u/\partial \sigma_x}{\partial u/\partial r_x} = \frac{r_m - r_f}{\sigma_m} \tag{1.2}$$

Equation (1.2) says that utility is maximized when the marginal rate of substitution between risk and return – the left hand side of the equation – is equal to the price of risk – the right hand side.

Define β_i as the risk of asset i in relation to the risk of the market as a whole. This allows us to express the total amount of risk in this asset as the product of β_i and the market risk: $\beta_i\sigma_m$. The cost of this risk is simply the price of risk as in (1.2) times this quantity or

$$\beta_i\left(r_m - r_f\right)$$

In equilibrium, the returns from holding any given asset, adjusted for the risk involved in holding the asset, must be equal to that for any other:

$$r_i - \beta_i\left(r_m - r_f\right) = r_k - \beta_k\left(r_m - r_f\right)$$

In particular, we may choose asset k to be the risk free asset so that β_k is zero:

$$r_f - \beta_f\left(r_m - r_f\right) = r_f$$

Rearranging, we get the basic equation of the capital asset pricing model, *CAPM*:

$$r_i = r_f + \beta_i \left(r_m - r_f \right) \tag{1.3}$$

Equation (1.3) is used with some of the models presented below. One more step, however, is necessary to produce the market line which will lie behind other parts of our empirical work. First, let us rewrite the equation as

$$r_i = r_f (1 - \beta_i) + \beta_i r_m$$

Letting $r_f (1 - \beta_i) \equiv a_i$ we may rewrite (1.3) in a form that is known as the *single index model*:

$$r_i = a_i + \beta_i r_m \tag{1.4}$$

Here a_i represents that component of the return to asset i that is independent of the return r_m to the market portfolio and thus may be reduced by the diversifying ones portfolio. a_i is really a random variable. Part of it is determined by the risk free rate of return, but the other part may be thought of as a purely random phenomenon. It is thus helpful if we write $a_i = \alpha_i + \varepsilon_i$ and consider ε_i as random and α_i as a deterministic function of the risk free rate of return. Thus the *market line* may be written as

$$r_i = \alpha_i + \beta_i r_m + \varepsilon_i \tag{1.5}$$

β_i itself cannot be observed, but may be estimated by statistical methods – using ordinary least squares (*OLS*), β_i will be the ratio of the covariance between the returns to asset i and the market portfolio and the variance of the market portfolio,

$$\beta_i = \frac{\text{cov}(r_i, r_m)}{\text{var}(r_m)}$$

We are now in a position to estimate the coefficient β_i for the stocks on the Stockholm Exchange. However, before we proceed, I will digress and give a short review of the history of this regression coefficient. Particular note will be made of the assumption of $beta_i$'s stability.

1.3. A History of Beta

The subject of this book is the question of the constancy of regression coefficients. We will examine a variety of other subjects, of course, but we will always return to this basic question.

The coefficient that will be used to illustrate the points made in this book will be the *beta* coefficient familiar to all who has studied Finance.

Thus it seems appropriate that we begin by tracing the history of this coefficient in the literature.[4] We shall be studying a rather mundane coefficient that has grown far beyond its rather humble origins as a parameter in an ordinary least squares regression equation. It is one of the few regression coefficients, simple or otherwise, that people actually pay money to get. As a concept in the theory of Finance, it has been around since the late 1950's. Markowitz introduced the concept if not the name as a practical tool to analyze returns to different stocks in relation to the returns to a market index. (Markowitz (1959), p. 100) He suggested that investors hypothesized a linear relationship between the return r_a to holding a given asset a and the return to holding the market portfolio I. The equation in his footnote, $r_a = A_1 + A_2 I + u$, with u being the random deviation between these two returns, is what later became know as the *market line*. The next reference to this line is found in Sharpe in his 1964 article in the *Journal of Finance*, but perhaps Jack Treynor in an unpublished and really undated manuscript predates Sharpe.

I quote Sharpe, vintage 1964, who uses g to refer to the market portfolio:

"[The] economic meaning can best be seen if the relationship between the return to asset i and that of combination g is viewed in a manner similar to regression analysis. Imagine that we were given a number of (*ex post*) observations of the return to the two investments. ... The scatter of the R_i observation around the mean (which will approximate E_{Ri}) is, of course, evidence of the total risk of the asset – σ_{Ri}. But part of the scatter is due to an underlying relationship with the return on combination g shown by B_{ig}, the slope of the regression line. The response of R_i to changes in R_g (and variations in g itself) account for much of the variation in R_i. It is this component of the asset's total risk which we will term systematic risk." (Sharpe, 1964, pp. 438-39)

Referring to Markowitz, Litner (1965) also writes out the market equation explicitly as $r_i = a_i + b_i I + u_i$ where, again, r_i and I are, respectively, the return to asset i and to a market index and u_i is a random deviation.

Note that this *market line* is a relationship that is supposed to hold for a single period. It is not developed directly from an equilibrium model or from the capital assets pricing model based on expectations rather than *ex post* observations. Indeed, it is not difficult to show that this line and the assumptions necessary for its existence as an equilibrium model are

---

[4]This survey is brief and not all inclusive. It is limited above all by the availability of the relevant journals in our library in Lund. Periodicals such as the *Financial Analyst's Journal*, the *Journal of Business, Finance and Accounting* and the *Journal of Economics and Business* are not available. Even some issues of the *Journal of Finance* and the *Journal of Financial and Quantitative Analysis* have either "walked away" or have not been purchased by the library.

inconsistent. (Fama (1968), p. 39). To be valid, the market model requires independence between the market index and the error terms u_i. But by definition, $I = \sum x_j r_j$ where x_j is the weight of asset j in the portfolio and the summation is over all assets. Obviously, the variance of r_i in the market equation (1.5) will contain a term for the covariance of I and u_i. From the definition of the returns to the market index, it is apparent that this covariance is not zero as the covariance between r_i and u_i is non-zero. Fama points out that this really does not matter, as one can substitute another index, say r_m which is a factor common to all assets on the market, for the returns to the market index in the market line and that the systematic risk will be approximately the same as before.

Fama's article is notable for an entirely different reason: it is the first one to use the symbol β for the systematic risk. Indeed, in Finance, these two terms have become synonymous.

The proponents of the market line assumed that returns were Gaussian. Another topic of debate during the 1960's was whether this assumption could be justified. Mandelbrot (1963, 1966), Fama (1965), Sharpe (1964), and Fama *et al.* (1969) were amongst the participants in this discussion. The alternative to the Gaussian model was the family of stable distributions — of which the Gaussian is a special case. The main difference between the models considered was that the Gaussian has a finite second moment while the stable distributions studied did not. In an article in the *Journal of Business* in 1969, Jensen discussed the implications of this debate for the capital asset pricing model. He shows that the systematic risk is more or less the same no matter which distribution returns actually follow (Jensen (1966), p. 206).

There is however, another problem to surmount: how can β be estimated? The market model is as it stands valid only for a single period. As the systematic risk varies among assets, this implies that we have but one observation available with which to estimated the coefficient, unless, of course, systematic risk is stable through time. Jensen addressed himself to this issue by expanding the single period model to a multi-period one which he called the "horizon solution". Again, I quote:

"One of the most important implications of the horizon solution given ... the restatement of the capital asset pricing results ... is the fact that the measure of the systematic risk, b_j, may be used for a holding period of any length, N. That is, we shall ... show that as long as H, the 'market horizon', is instantaneous, the expected value of the estimate of systematic risk ... will be independent of the length of time ... over which the sample returns are calculated and will be equal to the true value b_j. This result, of course, implies that we can use a given measure of risk for a portfolio's performance over a horizon of any length. In addition it means that we need only concern ourselves with the problem of obtaining the best estimate of b_j, for any security or portfolio, and any investor, regardless of his decision horizon, will be able to use that measure in arriving at an optimal portfolio." (Jensen (1966), p. 189)

What Jensen says is strictly true only if β_j, is stable. This is the next question which he addressed. Assuming that systematic risk is stable, then estimating the coefficient in two different time periods would give more or less the same result. Using mutual funds for the period 1945–1964, Jensen estimated b_j for the periods 1945–1954 and 1955–1964. He found that the correlation between the systematic risk of the funds during the two periods was 0.74. (Jensen (1966), p. 217) He also found that the number of negative changes in the coefficient lay within a confidence interval of two standard deviations (he used a normal approximation of a binomial distribution). His conclusion was that β_j was indeed stable.

1.3.1. IS BETA STABLE?

If the definition of systematic risk and the statement of the capital assets pricing model was the topic of the late 1960's the question of its stability was to occupy the profession, to a greater or lesser extent, for the following two decades. We might also note, that the coefficient of systematic risk began to be called *beta* – an article by Blume in the *Journal of Finance* in 1971 (Blume, 1971) seems to be the trend setter. How, then, did one test for *beta* stability and what conclusions were drawn?

One such test has already been mentioned – Jensen's comparison of *beta* for two adjacent ten-year periods. Note that the choice of ten rather than say twelve or six years was arbitrary. Blume (1971, 1975) examined a larger number of assets and a shorter time span – seven years. He justifies this seven year period by noting that using a much shorter period such as two or three years introduces measurement error as manifested by the estimation of negative *betas* for individual assets. (Blume (1971), p. 6). To be explicit, for each seven year period Blume estimates the following regression equation for each of his assets:

$$r_{it} = \alpha_i + \beta_i r_{mt} + \varepsilon_{it} \tag{1.6}$$

r_{it}, r_{mt} and the error term ε_{it} are as above but now with a time dimension. Blume then builds portfolios based on the *beta* estimate for the individual assets. The n stocks with the n lowest *beta* were placed in one portfolio, the n stocks with the next n lowest *beta* were placed in a second portfolio and so on until all the stocks were placed in portfolios of similar risk. He then averages *beta* for each portfolio. Next he estimates *beta* for the individual assets in the next period and once again calculates the average *beta* for the portfolios formed using the first period's estimated coefficients. He then calculates the rank correlation between the portfolios' average *beta*. He repeats this procedure for four periods – build portfolios in one period and compare them with the risk of the same portfolio in the following period. He also varies the portfolio size from one to 100 assets. He finds the correlation to range from 0.62 for single stocks to close to 1.0 for portfolios of 50 stocks.

Rank correlation between portfolios is high, but the individual assets' *beta* changes between periods. Blume tries to capture this change by estimating the following equation for each pair of adjacent periods:

$$\hat{\beta}_{i,k+1} = a + b\hat{\beta}_{i,k} + e_k \qquad (1.7)$$

where the hat " ˆ " denotes the estimated value and k is the period — he uses six of these. According to Blume, the estimated coefficient b shows the tendency of *beta* to revert to its mean value — unity — between the periods. Thus, for *beta* less than one, the next period's *beta* will lie closer to one and vice versa. He also interprets an estimate of b that is itself less than one to imply that all *betas* do tend to a grand mean of unity and deviations from this mean are but temporary.

I find this reasoning rather strange. If *beta* is stable — as is required if the capital assets pricing model is to be a guide for investors — it should not tend to go anywhere. And while temporary fluctuations will of course be observed when returns are measured *ex post*, if *beta* tends to go anywhere it should be back to its own mean and not to a grand mean of unity. Secondly, if one takes (1.7) seriously, then running the regression backwards should yield an estimated coefficient equal to the inverse of β. Typically, this does not happen. The reason herein is not difficult to see. The variance of the estimated *beta*'s in one period will be much the same as the variance of the *beta*'s estimated for the preceding or for the following period. Thus for the regression coefficient in (1.7) we should — and indeed do — find that if

$$\frac{\mathrm{cov}(\hat{\beta}_{i,k},\hat{\beta}_{i,k-1})}{\mathrm{var}(\hat{\beta}_{i,k})} < 1 \qquad \cdot$$

then

$$\frac{\mathrm{cov}(\hat{\beta}_{i,k},\hat{\beta}_{i,k-1})}{\mathrm{var}(\hat{\beta}_{i,k+1})} < 1$$

as the denominators are about the same.

In passing, we should also note that Klemkosky & Martin (1975) compare the accuracy of *beta* forecasts using Blume's method of estimating an transition equation to the adjustment process used by Merril Lynch, Pierce,

Fenner and Smith, Inc., and to that suggested by Vasicek (1973) which employs a Bayesian updating formula. Merril Lynch *et al.* adjust *beta* by adding a fraction of the difference between the estimated *beta* and one to unity: this adjusts *beta* towards the "grand mean" of one which Blume assumes. Vasicek adjusts *beta* by first dividing it by the variance of its estimate and adding this to the "cross section" average of all *betas* for that period divided by the cross section variance. This sum is in turn divided by the sum of the inverse of the cross section variance and the inverse of the variance in the estimated of the *beta* to be adjusted. Klemkosky & Martin estimate four adjacent five year periods beginning with July 1947–June 1952 and ending with July 1967–June 1972. This allows them to forecast three periods as the estimation process requires two whole periods. Thus, using one period's *beta* to forecast the next period's, they find that for July 1957–June 1962 — estimated using the relationships obtained from the previous two periods — that Merril Lynch's method gives the lowest forecasting errors. However, for the two following periods, the Bayesian updating formula proved superior. They also observe that all three methods of updating *beta* gave a better forecast than the assumption of stability.

But I am digressing: the real message in Blume's — and in Klemkosky and Martin's — articles is that *beta* is not stable. I will trace the main points of the debate on *beta's* stability chronologically and at the same time try to distinguish two lines of approach. One such line follows the methodology of Blume: *beta* is estimated for given time intervals and then the estimates for adjacent periods is examined. A second line of approach uses or modifies existing statistical tests in examining *beta's* stability.

1.3.2. TRANSITION MATRICES

Blume suggested that *betas* be adjusted for changes between periods. One natural question to ask was what constituted a period when *beta* could be considered constant. Blume used a seven–year period. Gonedes (1973) found that this period was "optimal". Baesel (1974) estimated for different intervals — one year, two years, four years, six years and finally nine years of monthly data — and found that *beta* stability as evidenced in transition matrices increased as the estimation interval increased. An *s*–period "transition" matrix is a matrix showing the probability that *beta* in period $t + s$ belongs to risk class j given that it was in risk class i in period t. Risk classification is based on a ranking of individual *beta's* estimated in each period. Baesel used five risk classes and found that the probability that a stock was in the same risk class in two adjacent periods was rather low. In fact, it was highest — but still low — for the most and the least risky portfolios (0.34 and 0.33 respectively). The probability was only 0.23 that a stock in risk class 2 (the next lowest) remained in the same class until the following period. This data is from the one year periods, but the same tendencies are present for the longer periods as well. A similar transition matrix for the Stockholm Exchange shows these same tendencies although the probability to remain in the same class given two adjacent periods is larger as the *beta*'s were calculated using partially overlapping time periods (Hansson & Wells, 1987). Using a *chi-squared* test of the null hypothesis that all the frequencies were equal, Baesel concludes that "security *betas* are unstable" (Baesel, 1974, p. 1493).

Altman *et al.* (1974) use French weekly data on stock returns to calculate transition matrices. They used periods of 13, 26, 52 and 208 weeks in their work and found that the longer the estimation period, the higher the period to period correlation which is by in large comparable to similar studies that used US data. Roenfeld *et al.* (1978) add a new dimension to the study of transition matrices by estimating an initial period of given length — four years of monthly data — and calculating transition matrices for following periods of one, two, three and four years of monthly data. Again, they found that *beta* was more stable the longer the time period used. Their data was from the New York Stock Exchange.

Alexander & Chervany (1980) present transition matrices based on *beta*'s calculated for five different time intervals – one, two, four, six and nine years of monthly observations from the New York Stock Exchange. As in Baesel's and Hansson & Wells' studies, the probability in remaining in an extreme high or low risk pentad between two adjacent periods is greater than that of remaining in the same risk group for the other pentads. However, with periods of one, two and four years this difference is small indeed and of doubtful significance — but they do not perform significance tests on the percentage distributions. The difference is more pronounced in the six and nine year samples. The probability of remaining in the lowest pentad is 0.3 in the six year one, although the probability of remaining in the fifth is less than that of remaining in the second. For the nine year period, the probability of remaining in the high risk group is 0.44 and in the low one is 0.28. Thus the longer the period of estimation, the more *beta* tends to remain in the same pentad. Strangely enough, despite the evidence in their own transition matrix indicating stability in the extremes, they criticize Baesel's findings by noting that the mean absolute deviations of *beta* between adjacent periods are greatest in the two extreme pentads indicating of course greater variations in the coefficients. They also criticize Baesel's findings that the longer the period of estimation, the more stable *beta* becomes. However, here their criticism is unjustified as they also find evidence of increasing stability unless they remove the extreme pentads (which Baesel claims to be more stable than the others).

Theobald (1981) examined the alleged increase in *beta* stability with increasing sample size using some rather simple theoretical considerations. He develops — with the help of some simplifying assumptions that he later tests — exact correlation coefficients between *betas* in adjacent periods. While agreeing with the above reported studies that this correlation tends to increase with an increasing estimation period length, Theobald finds a limit to this increase. He concludes that correlation is greatest with a period of a bit more than eighteen years — 220 months to be exact — although the observed benefit in increasing the sample beyond ten years is limited. He used British data.

One reason suggested (see, for example, Alexander & Chervany (1980), p.128) for *beta* instability is *measurement error* — the theoretical *beta* relates *ex ante* expectations while estimated *beta* relates *ex post* observations. Scott & Brown (1980) claim that this type of error combined with autocorrelation in the residuals would result in unstable estimates of a stable coefficient. Changes in *beta* between periods would thus be spurious. Perhaps, but I find no evidence of auto-correlation in the returns to the Stockholm Exchange.

Francis (1975) has attempted to find general lead/lag relationships in the market model. If such relationships existed, then estimating the traditional market line (1.5) rather than an equation without these terms could introduce a bias in the estimates that in turn could result in different estimates for *beta* for different time periods. He found, however, no consistent lead/lag relationship.

A third reason for *beta* instability could be that the market reacts differently during bull and bear periods. This would yield different *betas* for different periods even if the coefficient had a stable bull and bear value unless of course the periods studied had a similar dynamic development. Longer periods would more likely have approximately the same number of bull and bear observations than shorter ones: thus the tendency to *beta* stability in longer periods. Kon & Jen (1978) estimated *betas* from bull and bear markets where the switch between each was determined endogenously via a "switching regression" technique. This technique divides the sample into two (or more) subsamples by seeking a switching point for the regression line during the sample studied. Kon & Jen (1978, 1979) also used switching regression to find shift points for mutual fund's *beta*. Using an alternative hypothesis of two regimes, they found that 32 of the 49 funds they studied rejected the null hypothesis of one regime at the 10% level. (Kon & Jen, 1978) Subsequently Kon & Jen (1979) they reexamine the 49 mutual funds and reject the null hypothesis of stability for 39 of the funds with the alternative of two regimes and for 14 of the funds with the alternative of three.[5] They write that they attributed "any *beta* nonstationarity to the timing activity of fund managers." (Kon & Jen (1979), p. 285).

Fabozzi & Francis (1979) tested for explicit differences in bull/bear *beta*'s but found what they called a perverse relationship: *beta* tended to decrease in bull periods which is not what one would expect. They also used data for mutual funds and conclude that their results "suggest that mutual fund managers have not outguessed the market. When statistically significant shifts were observed in the funds *beta* , the change was generally perverse." (Fabozzi & Francis, 1979, p. 1249)

Miller & Gessis (1980) take the partitioning of the sample one step further. Using a method called "partitioned" regression, they search the sample for subperiods that exhibit differing regression coefficients. These periods are identified by minimizing the total residual sum of squares for the various regressions. Of the 28 mutual funds they studied, only one was considered stationary — that is, the partition covered the entire sample.

1.3.3. FINDING A MODEL FOR BETA

If *beta* is not stable — and this is really the consensus — the natural question to ask is whether one can find a model to describe it. Write the market equation as

$$r_{it} = \alpha_{it} + \beta_{it} r_{mt} + \varepsilon_{et} \tag{1.8}$$

The difference between (1.8) and (1.6) is that both coefficients here are assumed to be time–varying. In the pages that follow, I will concentrate on *beta* as a time–varying parameter. *alpha* may of course also have a time dimension, but for the moment it is assumed to be constant. In a paper dating from the Budapest meetings of the Econometric Society in 1972, Schaefer *et al.* (1975) consider four different models for *beta*: the *OLS* models already mentioned, the random coefficients model (*RCF*), the random walk model(*RW*) and the mean reverting model (*MRV*). In the first of these, *beta* is a constant for all periods; in the second,

$$\beta_{it} = \bar{\beta}_i + \mu_{it} \tag{1.9}$$

and the coefficient fluctuates randomly about a mean value. Note that this model is known by various names, including "the dispersed coefficient model" as Schaefer *et al.* call it and "mean reverting model" as with μ_{it} being a random variable following a Gaussian distribution with a zero mean and a fixed variance, the coefficient tends to return towards its mean. In this book this model is referred to as the random coefficients model.

---

[5]Interestingly enough, two of the funds rejected against the alternative of three regimes but not against that for two.

The third of the four models is the RW model, which may be written as

$$\beta_{it} = \beta_{it-1} + \nu_{it} \tag{1.10}$$

where ν_{it} is Gaussian with a zero mean and a fixed variance. Here the coefficient wanders aimlessly around. Schaefer *et al.* cite two earlier studies of *beta* that have used this model as the starting point: Kantor in the *Financial Analyst's Journal* in 1971 and Fisher & Kamin in a 1971 paper presented at Chicago. Szeto (1973) also used this model, although the noise parameter ν_{it} was dated at $t-1$ instead of t as in (1.10).[6]

Finally, the MRV model is presented as

$$\beta_{it} = \phi\beta_{it-1} + (1-\phi)\bar{\beta}_i + \zeta_{it} \tag{1.11}$$

where, again, ζ_{it} is Gaussian with a zero mean and a fixed variance. In this model — sometimes called the return to normality model — *beta* moves towards its mean gradually. If ϕ is restricted to lie between zero and one, this equation may be given an interpretation similar to that for the well known Koyck lag: next period's *beta* will be a weighted average of this period's coefficient and its mean value. The number of periods required for say 50% of the adjustment towards the mean to be accomplished is $0.5/\ln\phi$. I should note also that Rosenberg (1973) suggests the MRV model for *beta*.

Note that all four models can be thought of as special cases of the MRV: the OLS model obtains when $\text{var}(\zeta_{it}) = 0$; the RCF when $\phi = 0$; and the RW when $\phi = 1$ Schaefer *et al.* present a method for distinguishing between the four models and conclude that *beta* is not stable and is best described by the MRV model over the 45 year period 1926-1971 using monthly returns studied. However, the merit in their paper is the introduction of various models for *beta* rather than the test methodology used. The authors are well aware of these short-comings. Their test is, they write, "though unbiased, ... neither a maximum likelihood method nor is it in any other sense optimal." (Schaefer *et al.*, 1975, p. 155)

One of the first statistical tests applied to the market model was that for heteroscedasticity. If *beta* is non constant, then estimating a constant *beta* will cause the residual variance to be dependent on the returns to the market index. For example, suppose that the true *beta* model is not a constant but rather a RCF model. Substituting (1.9) into the market model (1.6) yields

$$r_{it} = \alpha_i + \bar{\beta}_i r_{mt} + \mu_{it} r_{mt} + \varepsilon_{it} \tag{1.12}$$

Taking variances and assuming that the returns to the market index are

[6]See footnote 1 in Chapter 4, page 76.

independent of *beta*'s random factor, we find

$$\text{var}(r_{it}) = \text{var}(r_{mt}) \cdot \text{var}(\mu_{it}) + \text{var}(\varepsilon_{it}) \qquad (1.13)$$

Thus the residual variance in the market model will be heteroscedastic. Using the other models gives similar results with of course other functional forms for the heteroscedasticity. Fama *et al.* (1969) were aware of the problem but concluded that it was not real problem. Their conclusion was based on the examination of scatter diagrams of the monthly return to 47 stocks from the New York Stock Exchange. As noted by Bey & Pinches (1980), Miller & Scholes (1972) as well as Brenner & Smidt (1977) studied similar data but drew the opposite conclusion. Brown (1977) — in a comment on a 1975 article by Martin & Klemkosky who found little evidence of heteroscedastic residuals — found that the null hypothesis of homoscedasticity could be rejected at the 5% confidence level for about 35% of the 683 stocks in his sample. Bey & Pinches also found non constant residual variance, but the exact percentages depended of the sample studied.

More recently, the question has been raised as to whether the residual variance follows and *ARCH* or a *GARCH* process. Using US data, Bollerslev *et al.* (1988) estimate a *GARCH* version of the market model for three assets (treasury bills, bonds and stocks). They conclude that the conditional covariance matrix for the assets is not constant over time (Bollerslev *et al.*, 1988, p. 127). Stenuis (1988) does a similar but univariate study for the returns to the Helsinki Stock Exchange. She found evidence in the nonstationarity of market risk. Hall *et al.* (1989) have studied value weighted sector portfolios — for the mechanical engineering, the financial, the electrical and the chemical firms on the London Stock Exchange. They estimate a modified market model where expected returns are conditioned upon information available at the end of the previous period and where the assets are specified to follow an *ARCH* process or a *GARCH* process. They prefer the latter specification. Note also that a "time–varying" *beta* is available in both studies as the covariance at each point in time between the market returns and the portfolio returns divided by the former.

Fabozzi & Francis (1978) published a study modeling *beta* as a random coefficient as in (1.12). They estimate the two unknown variances on the right hand side of (1.13) for 700 stocks on the New York Stock Exchange. If the $\text{var}(m_t)$ is found to be significantly different from zero, then they conclude that the *beta* for the stock studied may be modeled by (1.9). They use Theil's method to obtain estimates of these variances. First they subtract the means from the two variables — thus eliminating α_i from the equation — and estimate (1.6) using *OLS*. They then estimate

$$e_{it}^2 = P_t \sigma_\varepsilon^2 + Q_t \sigma_\mu^2 + f_{it} \qquad (1.14)$$

using *OLS*. The term on the left-hand side of the equation is the square of the *OLS* residuals from the market equation; the last term on the right-hand side is a residual factor. The other two terms are:[7]

$$P_t = 1 - \frac{r_{mt}^2}{\sum_t r_{mt}^2} \tag{1.15}$$

and

$$Q_t = r_{mt}^2 \left[1 - 2\frac{r_{mt}^2}{\sum_t r_{mt}^2} + \frac{\sum_t r_{mt}^4}{\left(\sum_t r_{mt}^2\right)^2} \right] \tag{1.16}$$

Recall that the variables are now defined as deviations from their respective means. From (1.14) we get estimates of the two variances. Call these two estimates $\hat{s}_{i\varepsilon}^2$ and $\hat{s}_{i\mu}^2$. The final estimates of the unknown variances are then obtained by solving the system

$$\begin{bmatrix} \hat{\sigma}_{i\varepsilon}^2 \\ \hat{\sigma}_{i\mu}^2 \end{bmatrix} = \begin{bmatrix} \sum_t \lambda_{it} P_t^2 & \sum_t \lambda_{it} P_t Q_t \\ \sum_t \lambda_{it} P_t Q_t & \sum_t \lambda_{it} Q_t^2 \end{bmatrix}^{-1} \begin{bmatrix} \sum_t \lambda_{it} P_t \hat{e}_{it}^2 \\ \sum_t \lambda_{it} Q_t \hat{e}_{it}^2 \end{bmatrix} \tag{1.17}$$

and where

$$\lambda_{it} = 0.5 \left(\hat{s}_{i\varepsilon}^2 P_t + \hat{s}_{i\mu}^2 Q_t \right)^{-2} \tag{1.18}$$

Note that all summations cover the entire data set. The standard error for the variances in (1.14) provide an estimate for the variance of the estimate of the true variance of μ_t. Fabozzi & Francis claim that a this may be tested using a student's t-test . This may seem incorrect as the t-test allows negative estimates and the parameter estimated here is a variance and thus inherently positive. However, the method itself is unrestricted and may well yield a negative estimate. Thus the use of the t-test is valid. Their estimation procedure yielded 382 positive estimates of $\text{var}(\mu_t)$ of which 103 were significantly different from zero at the 10% level and 57 at the 5% level. To describe the variance in *beta* for the individual stocks, they divide the square root of the estimated variance of μ_t by the *OLS* estimate of the mean of *beta*. Needless to say, they did this only for those assets with a positive variance estimate. For those estimates that were significant at the 10% level, this coefficient of variation lay between 0.5 and 1.0 for 74% of the stocks. The corresponding figure for 5% level was 66%. They noted that only in one case was the coefficient greater than 2. They conclude that "Sharp's intuitively pleasing definition of aggressive and defensive stocks as stocks with betas larger or smaller than unity has little practical value unless more exact methods of estimating betas are employed." (Fabozzi & Francis, 1978, p. 109)

---

[7]Note that in Fabozzi & Francis, formula (1.16) is incorrectly rendered. My assumption is that this error is simply a typographical one and that their results are calculated using the correct formula.

Fabozzi & Francis conclude the article by examining a number of the stocks that had negative estimates of var(μ_t). They restrict the coefficients in (1.14) to be positive by minimizing the sum of squared residuals in that equation while restricting the estimated parameters to be positive. They claim that the results were similar to those obtained from the unrestricted estimates. Alexander & Benson (1982) have criticized this restricted estimation procedure and reestimated the data using their improved method. Unlike Fabozzi & Francis, they split the data into six year intervals — 1960 to 1965 and 1966 to 1971. They found that the returns of 32 of the 683 stocks studies showed evidence at the 5% level of being generated by a *beta* following a *RCF* model. They found 44 were significant in the latter period. These percentages are slightly lower than those reported by Fabozzi and Francis.

Finally, Fabozzi *et al.* (1982) use the technique outlined above in a slightly different manner. I might also note that in this article, the typo mention above is "corrected" — that is, equation (1.16) appears as above. Here they question the ability of the market line to predict returns to a specific stock. If the market equation (1.6) is true, then by replacing the market index with the *OLS* estimate of *beta* for a given stock, and the dependent variable by the average return to that stock, then the parameters of the estimation of this equation using all the stocks should be the average value of the risk-free rate (the constant in the estimation) and the average return to the market index (the slope coefficient). They consider the six year period 1966 to 1971 which is then split into two equal three year periods. Using the method outlined above, but of course with the returns to the market index replace by the estimated beta coefficients and the returns to the individual stock replace by their average return, they find evidence that the *RCF* model is correct. They claim that their work "discloses a disturbing empirical result on the risk-return relationship. Employing a stochastic parameter regression model, we find the slope of the risk-return relationship varies randomly from one security to the next. Moreover, the observed randomness was substantial." (Fabozzi *et al.* (1982), p. 29).

Again I digress: the question of interest is whether *beta* can be modeled as a specific process rather than a constant. Fabozzi & Francis (1979) as well as Alexander & Benson (1982) accept the *RCF* model. Another model that was introduced in the literature during the 1970's was the random walk model. As mentioned above, the earliest reference is to Fisher and Kantor both in 1971. Szeto (1973) also used this model in the paper he presented to the *Joint Automatic Control Conference* in Columbus in 1973. His paper used the Kalman filter to generate a sequence of time–varying betas . His motivation for using non constant regression parameters was as follows:

"Although index models alleviate the difficulties of estimating parameters for the portfolio selection problem, they do not eliminate altogether such difficulties. An investor still faces the problem of having to choose ... [the model parameters]. Traditionally, analysts always invoked the time–invariance assumption on these parameters, and apply [*sic*] linear regression techniques to past data ... to obtain estimates .... The time–invariance assumption is very unrealistic ...; it fails to make use of the information embodied in the sequential nature of the data, and it fails to account for the dynamic variation over time of the parameters." (Szeto (1973), p. 303)

Szeto chose the random walk model as it allowed period to period fluctuation and contained the constant parameter specification as a special case. One suspects also that his training in automatic control contributed to his choice of models.

Sunder (1980) introduced a test for a random walk model in the spirit of Fabozzi & Francis (1978). Although this test is presented later on in the book, we will give a brief description of it here as well. Let X be a 2 by T matrix with ones in the first column and the returns to the market index in the second and define $M = I - X(X'X)^{-1}X'$ and m_{ij} as the typical element in the matrix. Using this terminology, (1.14) may be expressed as[8]

$$\hat{e}_t^2 = \sigma_\epsilon^2 m_{tt} + \sigma_\mu^2 \sum_{i=1}^{T} m_{ti}^2 r_{mt}^2 + w_t \qquad (1.19)$$

where as before the variable on the left-hand side is the *OLS* residual vector — note that the means have not been subtracted in Sunder's presentation. (1.19) may in turn be estimated by *OLS* and the *sigma* coefficients will be consistent estimators of the unobserved variances. This is the same estimate as used by Fabozzi & Francis except that the weighting procedure of (1.17)–(1.18) is omitted. When applied to my data, I found that the results from the two methods are almost identical.

Similarly, to estimate the variance in the random walk specification of beta as in (1.10), one applies *OLS* to

$$\hat{e}_t^2 = \sigma_\epsilon^2 m_{tt} + \sigma_\nu^2 \sum_{i=1}^{T} \sum_{j=1}^{T} m_{ti} m_{tj} r_{mi} r_{mj} \min\{i, j\} + w_t \qquad (1.20)$$

Once again, the *OLS* estimates are efficient estimates of the unknown variances. Of course, the problem remains that these estimates may well be negative.

---

[8]When no confusion may arise, I drop the index i in relation to a specific asset. Equation (1.19) and all those that follow are of course valid for just one asset.

Sunder estimates (1.20) using 600 monthly observations of 127 stocks for the period 1926-1975. He found considerable evidence of non-stationarity in *beta*: none of the estimates were negative and 88% of the estimates were "significant" at the 5% level. His findings are not really surprising. Few if any firms maintain the same risk profile over a span of fifty years. When he split the data into subsamples — he considers periods of 300, 150 and 75 months — the results are less convincing. For example, using a sample from January 1951 to June 1963, only 8% of the estimates have a "t–statistic" greater than 1.64 and 49% of the estimates of var(ν_t) are negative. To be fair, this subsample appears to be rather stable while the other three 150-month periods exhibit a much greater degree of instability.

Using two tests, Alexander *et al.* (1982) also investigate *beta* stability against the alternative that the random walk model is correct. One is the test suggested by Sunder and outlined above; the second is one developed by LaMotte & McWhorter (1978). and is based on the partition of the sum of squared *OLS* residuals using a orthonormal transformation of the X matrix given above. This test is presented in detail in the Chapter 2. It is also more difficult to implement as it requires the eigenvalues of a T by T matrix. Alexander *et al.* examined 67 mutual funds and found that for 22 — about 33% — they rejected the null hypothesis (at the 5% for the LaMotte–McWhorter test and at the 6% level for Sunder's test) of *beta* stability. Together with Simonds, LaMotte and McWhorter (Simonds *et al.* (1986)) have also applied their test to financial data, using 100 stocks on the New York Stock Exchange. Dividing the period 1951–1974 into three eight year subperiods and defining the constant term in (1.8) in different ways — to correspond to the market model and to Sharp's and Litner's description of the capital asset pricing model — they rejected the null hypothesis of stability at most 30% of the time (for the CAPM in period 1967–74) and at the least 11% of the time for the market model in 1959–66.

Garbade & Rentzler (1981) discuss the *RW* model, but their interest lies in estimation rather than testing. They show how one can estimate the unknown variances using a Kalman filter and the assumption of Gaussian error processes. Fisher & Kamin (1985) also discuss the random walk model and its estimation using a Kalman filter. However, their presentation also assumes that the error term in the market equation (1.8) has a hetero-scedastic variance. They relate this variance to the returns to the market index in a very theoretical manner. Their results are not published but are alleged to reduce the one month ahead forecast variance compared to the method of using the *OLS* estimate of the past 60 months to forecast next month's *beta*. The improvement is about a 40% reduction in the mean square error; if the *beta* to which their estimate is compared is adjusted by Blume's method, the improvement is still on the order of 30%.

The most general of the models (1.9)–(1.11) is the *MRV* model as the other two obtain by placing restrictions on the parameters in this model. As noted above, if $\phi = 0$, then the *MRV* collapses to the *RCF* model and with $\phi = 1$ it becomes the *RW* model.[9] Bos & Newbold (1984) test if $\text{var}(\zeta) = 0$ against the alternative that $\text{var}(\zeta) > 0$ and $\phi \neq 0$. They find little support for this model — they reject only in 8% of the 464 stocks on the New York Stock Exchange they studied covering the period January 1970 to December 1979. The test used was the Lagrange multiplier test. However, if they repeat the process without specifying ϕ they reject in 58.6% of the cases. The test used here is based on the work of Davis (1977) as the usual Lagrange multiplier test is not valid when a "nuisance" parameter is present. Watson & Engle (1985) also present the test which is based on the score vector — that is, the first derivative of the likelihood function. Watson & Engle note that their test results are similar to those obtained using the Breusch–Pagan test. Returning to Bos & Newbold's article, we should note that more than half — 53.9% to be exact — of the ϕ's estimated when the null hypothesis of constant coefficients was rejected were negative.[10]

Faff *et al.* (1992) use the *MRV* model as the alternative when they test for *beta* stationarity for Australian equities. The test they employed was a locally best invariant test using a Burr type XII distribution:

$$F(X) = 1 - (1 + X^c)^{-k} \tag{1.21}$$

for $X > 0$. The parameters c and k are also positive. They find that, using the discrete definition of returns that is employed in this book, 15.1% of the 159 stocks studied for 1978-1982 and 22.6% of the 310 in the 1983-1987 sample rejected the hypothesis of constant coefficients at the 5% level. These figures obtained when Faff *et al.* used an equally weighted market index. The corresponding figures for a value weighted index are 35.8% and 21.0%. They conclude that generally, "across all variations of our analysis a nontrivial degree of nonstationarity was evident." (Faff *et al.* (1992), p. 267) Brooks *et.al.* (1992) have examined the same data but with another test and then compare the results. Here the test is a point optimal invariant one based on unpublished work by one of the authors. Using continuously compounded data and a value weighted market index, Brooks *et.al.* find that the null hypothesis of stability rejects for 13.8% of the first sample and 13.9% for the second. These rejection rates compare to the 11.3% and 12.9% which obtain using the local best invariant test of Faff *et al.*[11]

---

[9]The *moving mean* model, (1.27)–(1.28) is even more general. See on pages 115–117.

[10]I note in passing that Bos and Newbold have a bit of trouble with addition. They state that the null hypothesis is rejected in 272 cases, but their Table 2 which presents the distribution of the sign of ϕ adds up to only 271 cases.

[11]There is a rather large discrepancy for the first period between Table 1 in Brooks

There is another version of the *MRV* model in the literature which combines mean reversion to a random mean for *beta*. This specification was introduced by Ohlson & Rosenberg (1982). They write the market model as

$$r_{it} = \alpha_i + \left(\bar{\beta}_i + \nu_{it}\right) r_{mt} + \delta_{it} r_{mt} + \varepsilon_{it} \tag{1.22}$$
$$= \alpha_i + \bar{\beta}_i r_{mt} + \delta_{it} r_{mt} + \omega_{it} \tag{1.23}$$

where

$$\delta_{it} \equiv \beta_i - \bar{\beta}_i = \phi_i \delta_{it-1} + \chi_{it} \tag{1.24}$$

and

$$\omega_{it} = r_{mt} \nu it + \varepsilon_{it} \tag{1.25}$$

The constant "true" mean of *beta*, $\bar{\beta}_i$, here perturbed by the random variable ν_{it} which has a zero mean and a variance of λ. The variance of the residual in the modified market equation (1.23 is now

$$\text{var}(\omega_{it}) = r_{mt}^2 \lambda + \sigma_{i\varepsilon}^2 \tag{1.26}$$

If $\lambda = 0$, then this model becomes the *MRV* presented above. Note that (1.24) closely resembles the specification found in Fisher & Kamin. If $\lambda > 0$, the variance of ω_{it} becomes heteroscedastic. However, this characteristic is induced by the tendency of the mean of *beta* to vary randomly about its "true" and is thus explained by the market model as formulated in (1.23). Using data from the New York Stock Exchange for the 50 year period January 1926 to December 1975, Ohlson & Rosenberg estimate *beta* for an equally weighted portfolio against a value weighted market index rather than for individual assets. They contend, that if one can reject *beta* stability at this aggregated level, then surely the *beta*s for the assets that make up the portfolio must vary even more. A second reason for their choice was to minimize "background noise" by choosing a portfolio that is highly correlated with the independent variable. One result of their choice is that the variation of *alpha* "disappears in the background noise unless it persists for hundreds of months. Thus the stochastic process for alpha is completely obscured." (Ohlson & Rosenberg (1982), p. 128) They find also that a second order autoregressive term in (1.24) does not improve the model. Their results are that the model given by (1.21)–(1.24) gives the best explanation of the data. It is interesting to note that they estimate

et.al. and Table 1 of Faff *et al.*. The former reports a rejection rate — at the 5% level — of 11.3% or 18 of the 159 assets. The latter reports a rejection rate — again at the 5% level — of 29.6% or 47 of the 159 assets. However, both papers quote test results for the same test on the same data. They do not, however, report a probability as to which is correct. I might add that basing an article on unpublished material makes life extremely difficult for other researchers who might show an interest in the methodology employed.

$\phi = 0.9875$ and find a significant difference between this model and one with the restriction $\phi = 1$ which is almost the RW model but the mean of *beta* is allowed to adjust randomly. They explain this difference as the one between a convergent model and a divergent one: even if convergence is slow — it takes about 55 months to get half-way home — it is faster than infinite divergence. However, this latter model has a significantly higher log likelihood than the RCF model. They do not estimate the pure RW model.

Collins *et al.* (1987) apply the above model to individual assets as well as portfolios of varying sizes using weekly rather than monthly returns with two sample lengths, 250 and 500 consecutive weeks randomly chosen from the period July 1962–December 1981. They use a data base of 500 randomly selected securities from the New York Stock Exchange to compare the above model with the RCF model and the constant coefficients one. Following Ohlson & Rosenberg, Collins *et al.* also restrict ϕ to be positive for every asset. They maintain that negative values of ϕ are "counterintuitive". *A priori*, they "expect that the major determinants of *beta* risk, that is, the firm's operating characteristics and financial structure, are likely to change gradually over time. This implies that *beta* coefficients would tend to change smoothly from period to period rather than oscillating back and forth , as would be the case if $\phi < 1$." (Collins *et al.* (1987), p. 438) If the "major determinants" of *beta* are the characteristics of the firm rather than the "animal spirits" of investors in the market or a combination of the two, then Collins *et al.*'s argument would be convincing. However, they do not substantiate their claim. Indeed, $0 < \phi < 1$ implies that investors never over react. Unfortunately, $-1 < \phi < 0$ implies that they always over adjust. Both positions seem to me to lack merit. As Collins *et al.* explain, the large number of negative ϕ's found by Bos & Newbold can well be due to the exclusion of the random variation in *beta*'s mean value which induces a downward bias in the estimation of ϕ. Further, if the true model is a RCF one but a MRV as in (1.11) is estimated, the autoregressive parameter ϕ, which is estimating the correlation between two adjacent *betas*, will center on its theoretical value of zero: some will be positive, some negative. Thus one may accept one model due to spurious correlations. This then together with Collins *et al.*'s discussion on bias seems to be a more reasonable justification in restricting ϕ to lie between zero and plus one. This is the course followed later on in the book.

Collins *et al.*'s results are informative. For the sample of 500 weeks, they reject the null hypothesis of constant coefficients when comparing the MRV as presented by Ohlson & Rosenberg in 55.2% of the cases using individual assets at the 95% level. Rejection frequencies were higher when portfolios were used: 75% for a portfolio of 10 stocks, 95.8% for one of 50 stocks and

99% for one of 100 stocks. Rejections rates were slightly higher when the RCF model used as the alternative. Using the RCF as the null hypothesis, the rejection rates were 10.4% for single assets, 15.2% for portfolios of 10, 30% for portfolios of 50 and 44.8% for portfolios of 100 securities. With the shorter period, the results were not as pronounced. In particular, the rejection rates with the RCF model as the null were quite small: 3.2%, 3.2%, 8% and 12.8% respectively.

Before leaving this subject, I should mention to studies of *beta* stability that lie closer to home. One of these is a study by Johan Knif of the Helsinki Stock Exchange using monthly data from January 1970 through December 1985. (Knif, 1988,1989). He examines returns using techniques based on the cumulated sum of squared residuals from recursive estimations and estimates the RW, the RCF and the MRV models for the Finnish data. At the 5% level, almost all of the 41 stocks studied show signs of parameter instability: if one accepts as evidence of parameter instability that the null hypothesis rejects in at least one test, then all 41 were unstable. (Knif (1988), p. 15.) Note that "parameter stability" here refers to variations in both *beta* — which was modeled as a random walk under the alternative hypothesis — and the variance of the disturbance in the market model. His estimations are presented for each security; however, no standard errors are printed making it difficult to interpret the tables. However, he does compare the coefficient of determination, R^2, between the different models. He finds that for 38 of the 39 stocks he models, that the RCF yields the highest R^2. If, instead of using the updated coefficients he uses the one step ahead predictions, then the OLS model is best 30% of the time, the RCF 26%, the RW 1% and the MRV 42%. The remaining 1% represents a hybrid model where the variance in the RW model is adjusted in an ad hoc manner. Note also that 14 of the 39 estimated ϕ's in the MRV model are negative. His results seem to favor the constant coefficient's version of the market equation to an extent greater than that which is published in the periodicals. The difference between his conclusions and those of Collins *et al.* is striking.

The second study is of the Stockholm Exchange, Wells (1990). Using general tests for heteroscedasticity on the data from January 1971 to December 1989, I found that the null hypothesis of homoscedasticity can be rejected at the 95% level in 77.5% of time for the 71 stocks studied. I also forecasted *beta* using Blume's method given above and found that the forecasted *beta* gave a better prediction than the naive assumption that *beta* would remain constant.

A third study is also by Wells (1994) and is really a preliminary version of this book. In it, 10 stocks plus a portfolio of all 10 are modeled using the techniques to be presented here. The main innovation of this paper is the introduction of a moving mean model where the mean value of *beta* develops as a random walk while the coefficient itself is allowed to vary randomly about the trend. To obtain this model, write (1.6) as

$$r_t = \begin{bmatrix} 1 & r_{mt} & 0 \end{bmatrix} \cdot \begin{bmatrix} \alpha_t \\ \beta_t \\ \bar{\beta}_t \end{bmatrix} + \varepsilon_t \tag{1.27}$$

This equation represents only one stock — therefore there is no subscript "i" attached to the returns of to the coefficients. The second part of the model is

$$\begin{bmatrix} \alpha_t \\ \beta_t \\ \bar{\beta}_t \end{bmatrix} = \begin{bmatrix} \phi_{11} & 0 & 0 \\ 0 & \phi_{22} & 1 - \phi_{22} \\ 0 & 0 & 1 \end{bmatrix} \begin{bmatrix} \alpha_{t-1} \\ \beta_{t-1} \\ \bar{\beta}_{t-1} \end{bmatrix} + \begin{bmatrix} 1 & 0 & 0 \\ 0 & 1 & 1 \\ 0 & 0 & 1 \end{bmatrix} \begin{bmatrix} \xi_{1t} \\ \xi_{2t} \\ \xi_{3t} \end{bmatrix} \tag{1.28}$$

If $\phi_{11} = 1$, then α_t is a random walk; if in addition, the variance of ξ_{1t} is zero, then α_t is a constant. If this be the case, the model should be respecified so that a constant *alpha* is included in (1.27). Also, the dimension of the state vector would be reduced by one. This would remove the first element of the vectors and the first row and column of the matrices in (1.28). This model is similar in spirit of equation (5.21) introduced by Lee & Chen (1982) but without the disadvantages of containing linear and quadratic trend terms.

1.4. Some final remarks

Most economists today would consider *beta* to be time–varying. This position would seem to be more reasonable than the opposite. After all, market conditions are fluid; today's leading firm may well be tomorrow's follower. However, even firms themselves change: they expand into new markets, at times with new products obtained through mergers or research and development. They leave other markets as demand there for their products falls. Assuming that the risk in holding a firm's stock be constant over a longer period of time would appear thus to be a rather heroic assumption. But why should this risk be constant even in the short run of one or two years? I can give no satisfactory answer other than the statistical methods used in estimating it require more than one or two observations. And even this answer begs the question.

Early writers such as Jensen assumed stability and sought evidence to support their position. It soon became apparent, however, that non constancy was the rule rather than the exception. Statistical tests were used to confirm this position. Some researchers found evidence of heteroscedasticity in the market model. This suggested time–varying regression coefficients.

Others tested for and found evidence of systematic variation in *beta*. By modeling this coefficient as a random one — equation (1.9) — or as a mean reverting one – equation (1.11) — or even as a random walk – equation (1.10), researchers implicitly embraced the hypothesis of a time–varying *beta*. Indeed, one of the questions asked in current research is which model best describes the time path of systematic risk.

With the exception of the final chapter, the remainder of this book is devoted the topic of time–varying *beta*. Using data from the Stockholm exchange, we will seek models for the *beta*'s belonging to firms on this exchange. First we test the data for parameter stability. Then, finding it lacking, we present methods for estimating non constant *beta* coefficients and then perform this estimation.

Chapter 2

Tests for parameter stability

2.1. Introduction

In this study, we consider the possibility that the market model as usually estimated is wrongly specified. The model is an equilibrium one showing the relationship between market risk and the risk of holding a certain stock. There is no explicit time variable in this model. The stability of the covariances through time is not asserted. It is only when the model is tested that the restriction of a time invariant *beta* is introduced. Constant coefficients of risk cannot adequately model a dynamic world.

The implication of this restriction is apparent. Assume that the true market model formulation at time t is

$$y_{it} = \alpha_{it} + \beta_{it}g_t + \varepsilon_{it} \tag{2.1}$$

where y_{it} is the return to a certain stock, g_t is the return to the market portfolio and ε_{it} is a random variable with a zero mean and constant variance of σ^2. As explained in connection with equation (1.5) above, the coefficient α_{it} is a function of the risk free rate of return and β_{it} is the covariance between the returns to the stock and the market portfolio — divided by the variance of the latter — at time t. Let us see what happens when this equation is estimated under the assumption of a time invariant coefficients, α_i and β_i.

Let $\hat{\alpha}_{it} = \alpha_{it} - \alpha_i$ and $\hat{\beta}_{it} = \beta_{it} - \beta_i$ where the new variables represent the difference between the "true" time–varying coefficient and the estimated constant one. Subtract the estimated equation, $\hat{y}_{it} = \alpha_i + \beta_i g_t$, from the true one in (2.1) and use the above definitions to find that

$$y_{it} - \hat{y}_{it} \equiv e_{it} = \hat{\alpha}_{it} + \varepsilon_{it} + \hat{\beta}_{it}g_t \tag{2.2}$$

The residual is composed of what may be considered random elements, $\hat{\alpha}_{it} + \varepsilon_{it}$, and a random coefficient multiplied with a deterministic factor,

$\hat{\beta}_{it}g_t$. The residual variance, assuming of course independence between the terms and the determinateness of g_t is

$$\text{var}(e_{it}) = \text{var}(\hat{\alpha}_{it}) + \text{var}(\hat{\varepsilon}_{it}) + g_t^2\text{var}(\hat{\beta}_{it}) \tag{2.3}$$

To put it another way, if a variable coefficient is estimated as a constant one, then the resultant residuals will be heteroscedastic. This in turn suggests a strategy for testing for the presence of variable coefficients: use one of the many tests for heteroscedastic disturbances that exist in the literature. The rejection of the null hypothesis of homoscedasticity would indicate that there may well be variable coefficients in the model.

While general tests for heteroscedasticity will suggest that regression parameters may be time–varying, they will not indicate the specific form for this variation. The test for heteroscedasticity will recommend the rejection of the null hypothesis whether the coefficient fluctuates randomly through time or if it develops as a random walk. There also exists in the literature a number of tests where the alternative hypothesis specifies a particular model. Sunder (1980) presents tests for both a random walk and a mean reverting formulation for an equation with a single regressor. Simonds *et al.* (1986) present a test for a random walk *beta* derived from earlier work LaMotte & McWhorter (1978). These tests use a constant parameter model as the null hypothesis; if this hypothesis rejected the alternative is explicit.

There is, however, a second set of parameters that must also be considered. These are the hyperparameters of the system: the variances of the estimation error and of the noise process that drives the variable coefficients as well as ϕ in the mean reverting model as stated in equation (1.11). These may well vary over the sample period and the question of their stability must also be considered. The usual tests for heteroscedasticity will return positive values for models with constant parameters but with shifts in the hyperparameters. However, two tests that seem to work rather well in identifying the latter shifts are the cumulative sum of squares and the moving sum of squares test. Indeed, I call these shifts *structural shifts* to distinguish them from time–varying regression coefficients. Of these two tests, the former is rather well known as it is included in textbooks in econometrics. In an appendix to this chapter I show how the table of significant levels presented in these texts may be expanded to cover situations with more than 200 degrees of freedom which is the largest sample covered by Durbin's 1969 table. The second test is not as widely used but is just as easy to calculate and — especially with larger samples — confidence intervals are more easily obtainable.

To assume that coefficients are variable is to commit the same error that those who assume constant coefficients commit: here we will let the data speak. If the tests reject the null hypothesis of constant coefficients, then the use of rather sophisticated methods to estimate time–varying parameters may be justified. Thus, in this chapter, the different tests will first be presented and then applied to the data at hand.[1]

2.2. Tests for Heteroscedasticity

The tests presented here fall into two categories: those which are based on the residuals from the *OLS* estimates and those which are based on the recursive residuals. As the model studied here is but a simple regression model, my discussion of the tests will be limited to the case with but one independent variable. Nothing presented here is new: the formulae are presented for the convenience of the reader who is referred to the literature for a deeper and more general analysis to the tests.

2.2.1. TESTS BASED ON THE OLS RESIDUALS

The basic equation tested in this section is (2.1) with $\alpha_{it} = \alpha_i$ and $\beta_{it} = \beta_i$.[2] With these restrictions, the equation becomes the *OLS* regression of a T by 2 matrix $Z = (1 \quad g_t)$ on y_t. Five tests are outlined below; most are very familiar to econometricians and are easy to calculate. White (1980) and Breusch & Pagan (1979) provide tests are traditional ones for heteroscedasticity and are presented even in elementary texts. Ploberger *et al.* (1989) is of more recent date and is more specifically designed to test for the stability of regression coefficients. The ARCH test Engle (1982) and the Ljung–Box *portmanteau* test are here interpreted as general tests on model efficiency. These last two tests may be performed on either the *OLS* residuals or on the recursive ones presented in Section 2.2.2.

2.2.1.1 White's Test

White's test examines the relationship between the squares of the *OLS* residuals and the cross products of the independent variables. To derive the test statistic, let the elements t, j of the matrix Ψ contain the j'th element of the lower triangle of the cross-product matrix $Z_t' Z_t$ and the vector $\bar{\Psi}_j$ contain the lower triangle of the average cross-product $Z'Z/T$.[3] Let also $\hat{\varepsilon}_t$ be the estimated residual and $\hat{\sigma}^2$ the estimated residual variance $\sum_{i=1}^{T} \hat{\varepsilon}_t^2$. If the "true" covariance matrix of ε_t is $V = T^{-1}\sigma_i^2 Z'Z$, then the difference

---

[1]GAUSS programs for calculating all of these tests are available in ASCII code from our *ftp* server as explained in Appendix B. See page 151 for details.

[2]The index for the individual stock, i, will be dropped in the following discussion. Thus the *OLS* coefficients are α and β.

[3]In the context of equation (2.1), Ψ has $K = 2$ columns, one for the independent variable and one for the constant. However, the test is applicable to equations with many independent variables.

between this true covariance and one estimated under the assumption of constancy, $\sigma_t^2 = \sigma^2$, is $V - \sigma^2(Z'Z/n)$. The lower triangle of this difference is collected in a column vector D:

$$D' = T^{-1} \sum_{i=1}^{T} \left(\hat{\varepsilon}_i^2 - \hat{\sigma}^2 \right) \circ \Psi \tag{2.4}$$

" $\circ$ " indicates *element by element* multiplication so that the t'th row of Ψ is multiplied by the difference $\hat{\sigma}_t^2 - \hat{\sigma}^2$. Note that the $\hat{\sigma}_t^2$ is estimated by the square of the residual ε_t.

Let $\hat{\delta}_\sigma$ represent the vector of differences between the estimated variance of ε_t and the average variance of all T residuals: $\hat{\delta}_\sigma = \hat{\varepsilon}_t^2 - \hat{\sigma}^2$. The average covariance of D is the matrix B:

$$B = T - 1 \left(\delta_\sigma \circ \Psi \right)' \Psi \tag{2.5}$$

The test statistic is, then

$$W = T D' B^{-1} D \tag{2.6}$$

Given the null hypothesis of a constant variance for ε_t, W is asymptotically $\chi^2_{K(K+1)/2}$.[4] There is one other item that must be mentioned: if the matrix Z of independent variables contains a constant, then the matrix B will be singular. Thus, before the statistic in (2.6) is calculated, the row and column of B and the element of D that corresponds to the constant must be deleted and the number of degrees of freedom reduced by one. When testing equation (2.1) as specified above, the number of degrees of freedom is $(K \times (K + 1))/2 = (2 \times (2 + 1))/2 = 2$.

There is an alternative formulation to the above test where the square of the *OLS* residuals — $\hat{\varepsilon}_{it}^2$ — is regressed on the columns of Ψ. The test statistic is simple: the multiple correlation coefficient from this regression is multiplied by the number of observations. The resulting statistic follows, given the null hypothesis of homoscedasticity coefficients, a *chi–squared* distribution with the number of degrees of freedom being equal to the number of regressors in the auxiliary regression. White's assumptions guarantee asymptotic normality which in turn allows the *chi–squared* distribution to be used. In the applications below, I use equation (2.6) to test for heteroscedasticity. Non constant regression coefficients are a common cause of a

---

[4]White 1980 presents the assumptions necessary for this statistic to exist. This test assumes that the expected values of the squares of $\hat{\varepsilon}_t$ and the independent variable x_t are bounded and that the average covariance matrices are non-singular. However, even the fourth moments and the fourth order cross products must also be bounded. These assumptions insure the invertibility of B as well as the asymptotic normality of the random variables estimated by $\hat{\delta}_\sigma$.

varying residual variance when the equation is constrained to have constant coefficients. Thus a rejection of the null hypothesis is taken as evidence that the coefficients in equation (2.1) are non constant.

2.2.1.2 Breusch–Pagan's Test

As with the above test, Breusch–Pagan's test utilizes the residuals from the *OLS* estimation of the (2.1) given constant regression parameters. However, this test uses the square of the *OLS* residual, $\hat{\varepsilon}_{it}$, divided by its estimated variance, as the dependent variable and only the square of the independent variable — plus a constant — as the regressor:

$$\frac{\hat{\varepsilon}_{it}^2}{\hat{\sigma}^2} = a_0 + a_1 g_t^2 + u_t \tag{2.7}$$

u_t is the residual in the auxiliary regression. One half of the explained sum of squares from the above regression is the test statistic. Given the null hypothesis of constant coefficients, the test statistic follows a *chi–squared* distribution with one degree of freedom. This test requires that the fourth moments of Z exist and that the error terms be identically distributed with a zero mean and a finite variance. However, even an assumption of the normality of the error term in (2.1) will of course suffice.

The difference between White's as stated in (2.6) and Breusch–Pagan's test is that Breusch–Pagan develop their test using an explicit expression for the heteroscedastic disturbance whereas White's test makes no such explicit assumption as it is based on the rather general specification using the moment matrix of the independent variables. Assume, for the moment, that the true model has time–varying parameters and that β_{it} develops as a random walk. Assume, also for the moment, that it has been estimated as a constant. The variance of the error term will now be

$$g_t^2 \sigma_\xi^2 + \sigma^2$$

where σ_ξ^2 is the variance of the random walk process. This example is similar to the one that Breusch–Pagan used in their derivation of their test statistic. Thus it would seem that this test is almost tailor made for the problem at hand. None the less, the White test may also be seen as a general test of model specification (see White (1980), pp. 823-824).

2.2.1.3 The ARCH test

We turn now to more general misspecficiation tests. One of these is the *ARCH* test introduced by Engle (1982). As mentioned elsewhere, a number of articles have claimed that security returns or even the term structure of interest rates follow a process where the variance exhibits AutoRegressive

Conditional Heteroscedasticity. (See, for example, Engle *et al.* (1987), Bollerslev *et al.* (1988) and Hall *et al.* (1989).) Where as the other two tests presented above assume that the variance of the *OLS* residual can be described by a function of the independent variable in the estimated equation, the *ARCH* test assumes that the variance of the stock return is a function of prior "shocks" or "surprises" to the system. This variance is thus time–varying. It follows that both the covariance between the return to the stock and the market portfolio as well as *beta* will also be time–varying. Thus the *ARCH* specification is an alternative to the one considered in this book. It is important that even the Swedish data is tested for the presence of *ARCH* tendencies.

To this end we rewrite (2.1) as

$$\varepsilon_t = y_t - z_t'\Gamma \tag{2.8}$$

where $z_t = (\begin{array}{cc} 1 & g_t \end{array})$ and $\Gamma = (\begin{array}{cc} \alpha & \beta \end{array})'$. Note that the regression coefficients α and β are not time–varying. y_t, the returns to the stock studied, is assumed to be normally distributed with a mean of $z_t'\Gamma$ and a variance h_t which is expressed as

$$h_t = a_0 + \sum_{k=1}^{L} a_k \varepsilon_{t-k}^2 \tag{2.9}$$

It is implicitly assumed that L past "shocks" are relevant for the variance.

The test statistic obtains by regressing the squared, estimated *OLS* residual on a constant and L lagged terms. The R^2 from this estimation, multiplied by the number of observations yields a statistic that, given a null distribution that all a_k's except the intercept are zero, has a *chi–squared* distribution with L degrees of freedom. The test for the presence of *ARCH* is similar to White's test; here, however, the right hand variables are the lagged values of the squared residuals rather than the moment matrix of the independent variables.

2.2.1.4   The Ljung–Box Test

Another test that examines the influence of past observations is the Ljung–Box test which is often called the *portmanteau* test. It examines the residuals for the presence of autocorrelation of a specified degree. Like the Durbin–Watson test, this test will be positive if there is significant autocorrelation; however, it will also be positive if the model is wrongly specified. In this case, the influence of omitted variables or a incorrectly specified functional form may yield a positive result. The test itself is well known and here presented only for the sake of completeness:

$$Q = (T(T+2)) \sum_{k=1}^{L} \frac{r_k^2}{T-k} \tag{2.10}$$

r_k^2 is the k'th autocorrelation of the *OLS* residual. Given the null hypothesis of no autocorrelation, Q follows a *chi–squared* distribution with L degrees of freedom.

2.2.1.5 The fluctuations test

This test is one for examining the constancy of regression parameters rather than a strict test for heteroscedasticity. Ploberger *et al.* (1989) base their test statistic on changes in the parameter vector calculated for increasing sample sizes rather than on the residuals from these estimations. The basic idea is simple enough: if there is "too much" variation between the final estimation — the one that uses the entire sample — and estimations using only part of the sample, then the null hypothesis of constant regression coefficients may not be true. The crux of the matter is to define "too much". To this end, write (2.1) as

$$y_t = z_t' \Gamma_t + \epsilon_t \tag{2.11}$$

where, as above, where $z_t = (1 \quad g_t)$. However $\Gamma_t = (z_t' z_t)^{-1} z_t' y_t = (\alpha_t \quad \beta_t)'$ is *time-varying* in contrast to (2.8) above.

At each time $t = k, \dots, T$ a new parameter vector is estimated. The test statistic is based on the coefficient whose deviation from its final is greatest:

$$S_T = \max_{t \in [k, T]} \left[\frac{t}{\hat{\sigma} T} \cdot \left\| (z_t' z_t)^{0.5} \left(\Gamma_{it} - \Gamma_{iT} \right) \right\| \right] \tag{2.12}$$

where $\|q\|$ denotes the maximum value of the vector q, $\hat{\sigma}^2$ is the residual variance for the entire sample and the maximum is of course taken from k, which is the number of estimated parameters, to T. Ploberger *et al.* also present the distribution function for this statistic as well as a table of critical values. The advantage with this test is that it examines exactly that aspect of the problem that we wish to test.

2.2.2. TESTS ON RECURSIVE RESIDUALS

A second series of tests has been applied to the recursive residuals obtained from the market model regressions. Whereas the method used in calculating the fluctuations test in Section 2.2.1.5 requires that the *OLS* regression co-efficients be calculated for each observation, the recursive residuals require but a single matrix inversion for the initial estimate: subsequent coefficients are updated using a simple recursive formula. There is an additional advantage to using the recursive residuals rather than the *OLS* residuals: the former, given the normality of the *OLS* disturbances, are orthogonal under the null hypothesis. This independence may then be used in the construction of tests based on these residuals.

While most of the modern, commercially available regression packages provide a routine for the calculating the recursive residuals, it is but a simple matter to write a routine that will calculate them. The formulae are presented below. For convenience, I use capital letters to indicate all observations up to and including the current date, while small letters represent only the current observation. Thus Z_t will include all observations of the independent variables up to and including that at time t while z_t will represent the observations only at time t. The same convention applies to the dependent variable y_t. The estimated coefficient vector at time t is b_t. The initial estimate of the coefficient vector b_0 is estimated from the first k observations where k is the number of parameters in the model.

$$b_{t-1} = \left(Z'_{t-1}Z_{t-1}\right)^{-1} Z_{t-1}Y_{t-1} \tag{2.13}$$

$$b_t = b_{t-1} + \left[\left(Z'_{t-1}Z_{t-1}\right)^{-1} z_t \left(y_t - z'_t b_{t-1}\right)\right]/f_t \tag{2.14}$$

$$\left(Z'_t Z_t\right)^{-1} = \left[\left(Z'_{t-1}Z_{t-1}\right)^{-1} z_t z'_t \left(Z'_{t-1}Z_{t-1}\right)^{-1}\right]/f_t \tag{2.15}$$

$$f_t = 1 + z'_t \left(Z'_{t-1}Z_{t-1}\right)^{-1} z_t \tag{2.16}$$

Note how "surprises" affect the system. The current observation on the independent variables is multiplied by the "historical" regression coefficient and the result subtracted from the current observation in (2.14):

$$v_t = y_t - z'_t b_{t-1} \tag{2.17}$$

The residual v_t is the recursive residual upon which the tests in this section are based; it is this residual that is the "prime mover" in the updating of the regression coefficient.

It should be noted that the recursive residuals are not uniquely defined. They depend upon the number of observations used to define the initial values of the regression coefficient b_0 and the cross-product matrix $\left(Z'_0 Z_0\right)^{-1}$. In the applications presented below, the number of observations used for this purpose is equal to the number of estimated coefficients, but this is no formal requirement. All that is needed is that the matrix be invertible.

The of the four tests presented below, two are parametric and two are really indicators of the adequacy of the model. The reader is referred to the discussion following the presentation of the tests for further comments on the different tests.

2.2.2.1 The von Neumann Test

This test, which is a variation on the Durbin–Watson test for recursive residuals, checks both first order autocorrelation and the specification in general. (Harvey (1990), chapter 5). The test statistic for the modified von

Neumann test is simple:

$$VN = \frac{T-k}{T-k-1} \cdot \frac{\sum_{t=k+2}^{T} \left(v_t - v_{t-1}\right)^2}{\sum_{t=k+1}^{T} v_t^2} \qquad (2.18)$$

Here k is the number of estimated parameters. Critical values for this statistic have been tabulated by Press and Brooks and are reprinted in rather many texts including Harvey (1990), Johnston (1984) and Ghosh (1991). However, if the sample is large, the critical values may be found using a normal approximation with a mean of 2 and a variance of $4/(T-k)$. Harvey suggests a one-sided test against the alternative of positive autocorrelation.

2.2.2.2   The Goldfeld–Quandt Test

This is a very well known test that hardly needs any special mention. However, as the test done below is slightly different from the standard version, I should point out the difference. In its more familiar form, one tests the hypothesis that residual variance is proportional to an independent variable. Here the independent variable is time: I test whether the volatility in the final L recursive residuals is the same as it is in the first L where $L \approx T/3$. Thus the F–statistic will have the same number of degrees of freedom in the numerator and the denominator which in turn implies that the test is statistic is symmetrical. Both large and small values will be significant as, *a priori*, the volatility may increase or decrease. The test statistic is as follows:

$$F(L, L) = \frac{\sum_{t=T-L+1}^{T} v_t^2}{\sum_{t=k+1}^{k+L} v_t^2} \qquad (2.19)$$

2.2.2.3   The Cumulative Sum of Squared Recursive Residuals

This test as is best thought of as an indicator of structural stability. It calculates the fraction of the total variance of the residuals up to and including the current observation. This fraction is then plotted for the entire sample. If the residuals are all about the same in absolute value, the plot is a more or less straight line from zero to one. Periods with greater — or lesser — volatility will show up as sharp divergences from this line. This method was introduced by Brown *et.al.* (1975) as an alternative to the simple cumulative sum of the recursive residuals and has become a more or less standard tool in the analysis of economic time–series data.[5]

[5]The cumulative sum of the recursive residuals appears to have been introduced by Page (1954); a related test, the "recursive t–test" has been suggested as a parametric alternative (see Harvey (1990), chapter 5). See Chapter 5 below.

The test statistic is the maximum value of

$$S_i = \frac{\sum_{t=k+1}^{i} v_t^2}{\sum_{t=k+1}^{T} v_t^2} \qquad (2.20)$$

where $i \in [k+1, T]$. Critical values for S_i are necessary to add content to the phrase "too great a deviation from the expected value". Given the null hypothesis of no structural change, S_i can be shown to follow a beta-distribution with a mean of $(i-k)/(T-k)$. The critical values can thus be expressed graphically as a pair of parallel lines

$$\pm c_0 + \frac{i-k}{T-k}$$

Durbin (1969) has tabulated c_0 under the assumption that S_i crosses either the upper or the lower boundary. The table is has been reprinted in a number of books on elementary econometrics and is thus readily available. In an appendix to this Chapter, this table is expanded to cover the length of the series considered here. (see Edgerton & Wells, 1994). It should be noted that Brown *et.al.* (1975) recommended that this test be used an indicator of structural change rather than as a method of exactly determining such change and the time of its occurrence. Thus the critical level should be chosen at the 10% level or higher as it is a greater mistake to accept the null hypothesis of no structural change when in fact there has been such a break than it is to reject the null hypothesis when there in fact has been none. Ploberger (1989) points out that this test has nontrivial local power against the alternatives where the variance processes of the system are time dependent.

The "qualitative" nature of this test has been criticized widely. For example, Hays & Upton (1986) reject the test as it only considers the case where S_i crosses the critical boundary once when, in practice, more than one crossing may occur. They recommend using Quandt's log-likelihood ratio technique which unfortunately has no known distribution thus making it impossible to compute any confidence level for the statistic.[6] Personally I prefer some if even vague critical area to none at all. It has also been suggested by Phillips in his discussion of the Brown *et.al.* paper that the S_i may be sensitive to departures from normality as there is some evidence

---

[6]The statistic is the log-likelihood ratio: $\lambda_t = 0.5t \ln(RSS_1/t) + 0.5(T-t)\ln(RSS_2/(T-t)) - 0.5T \ln(RSS_T/T)$ where RSS_1 is the residual sum of squares from the *OLS* regression of the first t observations, RSS_2 is that for the remaining $T-t$ observation and RSS_T is that for the entire sample. Where λ_t obtains a minimum value is a point where structure has changed. However, λ_t may have local as well as global minimum values and the lack of a critical value makes this test as arbitrary as the S_i statistic is accused of being. However, as it is not difficult to compute, one may compare its graph with that of S_i as an additional indicator of when a break may have occurred.

that the cumulative sum of recursive residuals test is not robust to such departures.

2.2.2.4 The Moving Sum of Squares Test

A less well known variant of the above test is based on the moving sum of squared recursive residuals. (Westlund & Törnkvist (1989)) The basic idea here is that the variance of v_t is calculated for a "window", w, of a given number of observations — say 6 or 12. The variance here is then compared to the variance of the entire sample with the current window removed from this calculated variance. The resulting statistic follows a F–distribution with $(w, T - k - w)$ degrees of freedom. Letting i take values between $k + w$ and T, the test statistic is the maximum of

$$M_i = \frac{\sum_{t=i-w+1}^{i} v_t^2}{\sum_{t=k+1}^{i-w} v_t^2 + \sum_{t=i+1}^{T} v_t^2} \cdot \frac{T - k - w}{w} \qquad (2.21)$$

Westlund & Törnkvist (1989) have examined the *mosumsq* test. They cite Hackl (1980) when presenting the confidence interval for the statistic. This interval is based upon Bonferroni's inequality and works for small probabilities, but not for larger ones. For example, when trying to find a probability for the maximum value of the *mosumsq* series when this maximum value lies under the confidence line yields the very unhelpful result that the probability is less than 1.5 or some other number greater than one. We knew this before we started. Thus only the maximum value and the 10% confidence bound has been presented. As above, one can plot M_i together with its 10% confidence line. If the former crosses the latter then one has an indication of a structural break. Hackl (1980) claims that this test has high power in identifying structural shifts either in variances or in coefficients.

2.3. Tests with a Specified Alternative Model

All of the above tests may be thought of as general tests for misspecficiation, although Breusch–Pagan's test is a likelihood ratio test where the alternative is completely specified. However, this specification may be both very specific as in Section 2.2.1.2 or rather general. The alternative in White's test, on the other hand, is rather general alternative to a homoscedastic model. The *ARCH* test is also rather vague in specifying the exact function form of the alternative model as it examines patterns in the residual variance rather than model structure. Below I present two tests based on explicit alternative models: the LaMotte–McWhorter and the Sunder tests.

2.3.1. THE LAMOTTE–MCWHORTER TEST

LaMotte & McWhorter (1978) present a test for a model whose coefficients, under the alternative hypothesis, develop as random walks. To be more exact, they consider the model

$$y_t = z_t'\beta_t + \varepsilon_t \tag{2.22}$$
$$\beta_t = \beta_{t-1} + \xi_t \tag{2.23}$$

While it is assumed that β_t and ξ_t are k-dimensional vectors and z_t is a $t \times k$ matrix — y_t and ε_t are scalars —, the variance of ξ_t is a multiple of a scalar σ_ξ^2 for each element of the vector. That is, the vector ξ_t is assumed to be normally distributed with a zero mean and a variance of $\sigma_\xi^2 I_T \otimes D$ where D is a known, diagonal matrix. Thus both σ_ε^2 — the variance of ε_t — and are scalars. We test whether the variance of ξ_t is zero — the null hypothesis — or positive — the alternative hypothesis. Under the null hypothesis, the regressors of equation (2.22) are constant. The test is rather involved and requires more computing power than do the of the tests presented above. My presentation follows LaMotte & McWhorter's article rather closely. By programming the equations in, say, GAUSS — LaMotte & McWhorter used SAS on a mainframe — the method is fully feasible to use on a PC with 8 megs internal memory.

Assume that the initial value of β_0 known. Then, by repeated substitution of (2.23) into (2.22), we find that

$$y_t = z_t\beta_0 + z_t \sum_{i=1}^{t} \xi_i + \varepsilon_t \tag{2.24}$$

Stacking the y_t's into a single T–dimensional vector and letting $Z = (z_1, z_2, \ldots, z_t)'$, equation (2.24) becomes

$$y = Z\beta_0 + \Psi\Omega + \varepsilon \tag{2.25}$$

where $\Psi = \text{diag}(z_t')$ and $\Omega = (\omega_1', \omega_2', \ldots, \omega_T')'$ and $\omega_i = \sum_{j=1}^{i} \xi_j$. This allows us to write the variance of y_t as

$$\text{cov}(y) = \sigma_\varepsilon^2 I + \sigma_\xi^2 G\,(I_T \otimes D)\,G' \tag{2.26}$$

where $G = (E_T \circ Z) \circ (L_T \otimes E_k)$. Here E_j is a $1 \times j$ vector of ones and L_j is a $j \times j$ matrix with ones on and below the diagonal and zeros elsewhere. As above, I use $\circ$ to denote element by element multiplication. The elements of $G\,(I_T \otimes D)\,G'$ may alternative be expressed as

$$V \equiv G\,(I_T \otimes D)\,G' \equiv \min\,(i,j) \sum_{s=1}^{k} d_{ss} z_{sj} z_{si} \tag{2.27}$$

The test statistic used is based on an invariant transformation of the dependent variable such that $e_t = H'y_t$. The matrix H will have the dimension of $T \times (T-k)$ and its columns will form an orthonormal basis for the vector subspace orthogonal to z_t. This implies that the following two relations hold:

$$H'H = I_{T-k} \tag{2.28}$$

$$HH' = I_T - Z\left(Z'Z\right)^{-1}Z' \tag{2.29}$$

Note that Z is a $T \times k$ matrix as in (2.25). That H is orthogonal to Z may be seen by post multiplying (2.29) by Z.

By construction $e'e$ is the residual sum of squares from the estimation of (2.22) under the null hypothesis. e is thus normally distributed with a zero mean and a variance of $\sigma_\varepsilon^2 I + \sigma_\xi^2 H'VH$. LaMotte & McWhorter show that the OLS residual variance may be written as

$$e'e = \sum_{i=1}^{T-k} Q_i = \sum_{i=1}^{T-k} e'P_i P_i' e \tag{2.30}$$

The columns of matrix P contain the normalized eigenvectors of $H'VH$ corresponding to the eigenvalues of that matrix arranged in descending order. Here I assume that the $T-k$ eigenvalues λ_i are distinct. As the statistic $\left(\sigma_\varepsilon^2 + \lambda_i \sigma_\xi^2\right)^{-1} Q_i$ follows a *chi–squared* distribution with one degree of freedom, an exact test of the null hypothesis is possible. We can partition the sum $e'e$ into two parts according to the sizes of the eigenvalues calculated above. The first part will contain enough terms to encompass 85% of the residual sum of squares. Call this number of terms n_1. Typically for the application at hand, this partition will contain three to five terms. This sum, divided by the number of terms, is, in turn, divided by the remainder of the sum divided by the number of remaining terms. The resulting statistic will, under the null hypothesis, follow an F–distribution with n_1 and $T-k-n_1$ degrees of freedom:

$$F(n_1, T-k-n_1) = \frac{\sum_{i=1}^{n_1} Q_i/n_1}{\left(z'z - \sum_{i=1}^{n_1} Q_i\right)/(T-k-n_1)} \tag{2.31}$$

2.3.2. THE SUNDER TEST

Another test may be run using the same data. The OLS residual variance, given the model with a constant and one random walk coefficient, may be decomposed into two components. The first of these is the variance of the error term in (2.22) multiplied by the diagonal elements in HH'; the second is the variance in the error term in (2.23) multiplied by the diagonal elements of V pre- and postmultiplied by HH', where V is defined by (2.27). Sunder (1980) suggested testing for a random walk *beta* against the alternative of constant coefficients by regressing the square of the residual from the *OLS* estimation of (2.22) on the diagonal elements of M and MVM where M is the matrix HH' defined in (2.29). The coefficients of the regression provide estimates of the variances of the error terms in the equations (2.22) and (2.23). (see Sunder, 1980) Let A_{ii} be the i'th diagonal of the matrix A. The regression equation may then be expressed as follows:

$$e_t^2 = \sigma_\varepsilon^2 M_{tt} + \sigma_\xi^2 [MVM]_{tt} + \zeta_t \qquad (2.32)$$

where ζ_t is the error term in the regression. A t-test against a one-sided alternative provides the test statistic. One problem does arise in that there is nothing in the estimation algorithm that guarantees that the variances be positive. However, if negative terms estimate, they are generally not significantly different from zero. That is, the confidence interval for the negative coefficient usually encompasses the origin.

Simonds *et al.* (1986) argue that their test has greater power than the test proposed by Sunder and that the power of Sunder's test is low when the variance of g_t in equation (2.1) is small when compared to the variance of ε_t. This relationship between the two variances seems to be the case for stock market data.

2.4. A Discussion of the Test Results

This section presents the results of the tests and compares them with previously published results. As test results vary greatly depending on the test used, a final section asks why this should be so.

2.4.1. THE RESULTS OF THE TESTS

Table 2.1 presents the *p-values* of the above tests when performed on the 21 of the stocks in the data base that were listed on the Stockholm exchange after 1 January 1991. The data are defined on page 92.

One of the first things one notices about the table is that the results are very mixed: for the most part the different tests yield different results.[7] Only in three of the 21 cases do all seven of the tests yield the same result: two — Garphyttan and SKF — with all seven rejecting the null hypothesis of constant parameters and one — Pharos — with all seven accepting this

---

[7]In the discussion that follows, I reject the null hypothesis at the 10% level. The sin of rejecting a true null is deemed to be less than that of accepting a false one.

hypothesis. There are three instances where six of seven tests reject — for Export and Investor, where the Fluctuations test does not reject the null hypothesis, and for Pharmacia, where the von Neumann does not reject; there is one case where six of the seven do not reject: only the Goldfeld–Quandt rejects for Åkermans. Continuing, there are two cases where five reject and four where five accept; in four instances four of the tests reject while four accept in four cases. [8]

There is no obvious pattern in the test results: thus the White test is significant in 13 of the 21 cases (62%), the Breusch–Pagan in 15 (81%), the fluctuations in 8 (38%), the *ARCH* in 10 (48%), the Ljung–Box in 7 (33%), the modified von Neumann also in 7 (33%) and the Goldfeld–Quandt in 17 (81%).[9] The White and Breusch–Pagan tests agree 72% of the time; these two and the *ARCH* agree 52% and these two and the Goldfeld–Quandt agree 71% of the time. When both of the first two reject, the *ARCH* test rejects 67% of the time (in 8 of 12 cases). The *ARCH* test and the Ljung–Box agree 81% of the time; the *ARCH* and the Goldfeld–Quandt 48%. Looking at the results from a different point of view, we see that if both the White and the Breusch–Pagan tests accept the null hypothesis, the *ARCH* rejects in 3 cases; where both of the first two reject, the Arch accepts in 4 cases.

If we expand our frame of reference to also include Table 2.2, the situation becomes a bit more definite: the tests for the random walk model seems to indicate that a large number of the stocks could be modeled in this way. One or both of the two tests applied are positive for 15 of the 21 stocks; in 4 of the cases, both tests are negative. For the remaining 2 stocks, the LaMotte–McWhorter test was negative but the Sunder test estimated a negative variance — but such estimates were not significant from zero. It is also interesting to note that in 12 cases where either the White or the Breusch–Pagan tests were significant then either the LaMotte–McWhorter or the Sunder test was also positive. The reverse — no to one or both of the former and no to one or the other of the latter — occurred 4 times. In 3 cases, both the former were negative and at least one of the latter was positive; and all four tests were negative only twice. It is also interesting to note that in both of these cases, either the *ARCH* or the Goldfeld–Quandt test was positive.

Taking the two tables together, we find that all of the 21 stocks reject the null hypothesis of constancy for at least one test.

---

[8]In all, there are 57 stocks in the database that were listed after 1990. All have been examined but the test results here, and in Chapter 5 the estimates, are presented for a very non random sample of 21. The conclusions drawn from the entire sample are about the same as above: in 3 cases, all tests agree; in 15 six of seven agree; in 21 five of seven; and in 18 four of seven.

[9]For all 57 stocks, the corresponding percentages are, in order, 25%, 44%, 30%, 28%, 26%, 26% and 53%.

TABLE 2.1. **Tests for constant parameters.** Except for the fluctuations test, the entries in the table are the probability that the null hypothesis is true.

Stock	T^a	White[b]	Pagan[c]	Fluct[d]	Arch[e]	L–B[f]	v–Neu[g]	G–Q[h]
Aga	252	0.500	0.027	2.472	0.372	0.012	0.002	0.149
Alfa Laval	247	0.000	0.008	0.554	0.667	0.498	0.115	0.015
Asea	252	0.000	0.001	1.291	0.006	0.169	0.204	0.004
BGB	104	0.465	0.531	2.012	0.003	0.065	0.995	0.911
Custos	251	0.372	0.092	1.275	0.125	0.678	0.546	0.000
Ericsson	252	0.001	0.012	1.338	0.978	0.518	0.935	0.000
Esab	252	0.931	0.394	0.760	0.000	0.060	0.038	0.404
Export	249	0.000	0.031	1.193	0.008	0.000	0.000	0.002
Gambro	104	0.518	0.209	1.565	0.976	0.395	0.220	0.014
Garphyttan	252	0.000	0.033	2.132	0.031	0.000	0.013	0.000
Industrivärden	251	0.051	0.052	1.230	0.317	0.045	0.264	0.013
Investor	251	0.004	0.099	0.577	0.048	0.024	0.018	0.000
Munksjö	238	0.000	0.008	2.069	0.000	0.241	0.519	0.055
Pharmacia	235	0.033	0.002	1.349	0.034	0.006	0.563	0.014
Pharos	131	0.864	0.102	1.607	0.735	0.206	0.792	0.846
Protorp	131	0.001	0.209	0.994	0.474	0.233	0.170	0.034
Providentia	251	0.009	0.046	0.498	0.000	0.084	0.072	0.000
SKF	252	0.003	0.023	2.115	0.019	0.000	0.001	0.000
Trelleborg	251	0.994	0.001	0.632	0.105	0.669	0.067	0.005
Volvo	252	0.037	0.005	1.895	0.925	0.361	0.932	0.055
Åkermans	239	0.123	0.347	0.967	0.851	0.989	0.166	0.000

[a]The number of observations.
[b]White's test (White, 1980).
[c]The Breusch–Pagan test (Breusch & Pagan, 1979).
[d]The fluctuation test: the 10% level, 1.35; the 5%, 1.48. (Ploberger *et al.*, 1989).
[e]The *ARCH* test (White, 1980).
[f]The Ljung–Box test.
[g]The von Neumann test (Harvey (1990), ch. 5).
[h]The Goldfeld–Quandt test.

Other comparisons may be made but the bottom line says that the tests results are different for the differing tests. Before inquiring into the reason for this diversity of results, we should compare Tables 2.1 and 2.2 with the results from other studies.

2.4.2. COMPARING PREVIOUS RESULTS

Martin & Klemkosky (1975) found little evidence of heteroscedasticity in the 355 stocks, for the period April 1967 to July 1973, on the New York Stock Exchange that they studied: their Goldfeld–Quandt test was signifi-

TABLE 2.2. **Tests for Random Walk beta.** n_1 and $T-k-n_1$ are the degrees of freedom in the numerator and the denominator of the LaMotte–McWhorter test.

Stock	n_1	df^a	LM^b	p–LM^c	var$(\varepsilon)^d$	var$(\xi)^e$	t–var$(\varepsilon)^f$	t–var$(\xi)^g$
Aga	4	246	3.926	0.0041	31.74	0.0021	8.566	1.286
Alfa Laval	4	241	1.688	0.1535	30.03	0.0054	6.708	2.785
Asea	4	246	3.129	0.0155	35.74	0.0036	5.610	1.285
BGB	4	98	1.561	0.1907	76.30	0.0443	3.130	2.237
Custos	4	245	0.768	0.5473	38.23	0.0001	7.182	0.063
Ericsson	4	246	0.625	0.6449	59.63	0.0035	5.067	0.684
Esab	4	246	0.807	0.5219	58.81	0.0026	6.532	0.647
Export	4	243	1.099	0.3577	23.47	0.0065	4.333	2.679
Gambro	4	98	0.726	0.5762	82.36	-0.0069	3.534	-0.365
Garphyttan	4	246	2.730	0.0298	101.09	0.0250	3.785	2.129
Industrivärden	4	245	4.312	0.0022	23.97	0.0034	6.616	2.076
Investor	4	245	4.088	0.0032	19.11	0.0045	5.400	2.851
Munksjö	4	232	3.547	0.0079	143.09	0.0187	5.055	1.515
Pharmacia	4	229	1.588	0.1783	50.66	0.0026	4.809	0.456
Pharos	3	126	2.181	0.0935	57.68	0.0040	4.919	0.658
Protorp	3	126	0.827	0.4815	27.12	0.0139	3.907	3.806
Providentia	4	245	3.437	0.0094	20.67	0.0046	6.541	3.247
SKF	4	246	3.617	0.0069	34.90	0.0063	6.618	2.708
Trelleborg	4	245	2.198	0.0698	77.20	-0.0040	7.433	-0.863
Volvo	4	246	2.352	0.0547	30.52	0.0032	6.841	1.617
Åkermans	4	233	2.246	0.0649	57.40	0.0040	2.832	0.450

[a]Degrees of freedom: $T - k - n_1$.
[b]The LaMotte–McWhorter test; see equation (2.31).
[c]p-value of the LaMotte–McWhorter test.
[d]The estimated variance in the Sunder test; see equation (2.32).
[e]The estimated variance in the Sunder test; see equation (2.32)
[f]The t–value for the estimate of var(ε).
[g]The t–value for the estimate of var(ξ).

cant on 8.4% of the time.[10] These findings were criticized by Brown (1977) indexBrown, S. J. who found a much greater degree of heteroscedasticity using 683 stocks for the period January 1961 to June 1968. He reported that the Goldfeld–Quandt test rejected the null hypothesis about 35% of the time. Morgon (1976) tests they hypothesis of normally distributed returns against an alternative hypothesis that the variance of the process is related to the volume of trading. Using a limited sample of forty stocks with monthly returns, he found that while the normal distribution is "not an adequate description of returns, it is sufficiently accurate to demonstrate the existence of heteroscedasticity." (Morgon (1976), p. 505) Bey &

[10]All the significance levels reported here are at the 5% level unless another level is explicitly stated.

Pinches (1980) examined the returns for 665 companies from January 1962 to December 1976. Dividing the period into three equal five year spans, they report that the Goldfeld–Quandt test rejects 21%, 24% and 33% of the time. As in the case with the tests reported in Table 1, they noted that different tests gave different results. Fisher & Kamin (1985) report that they have "found rather strong evidence that true residual returns were heteroscedastic." (Fisher & Kamin (1985), p. 132) Collins et al. (1987) report that the Breusch–Pagan test rejects the null hypothesis 30% of the time for a sample length of 250 and 60% of the time for a sample length of 600. They use 500 randomly chosen securities on both the New York and the American Stock Exchanges for the period July 1962 to December 1981.

The literature also contains some studies of heteroscedasticity non-US stock exchanges. Bey & Pinches (1980) cite Praetz's study of the Australian exchange in which he claims that returns there are heteroscedastic. They also cite Belkaoui's study of the Toronto Stock exchange. His Goldfeld–Quandt test shows that 40% of the returns are heteroscedastic at the 10% level. Knif (1989) found considerable heteroscedasticity on the Helsinki exchange; Wells (1990) reports similar results to those presented above for the Stockholm exchange.

Comparing the values in Table 2.1 with the above studies, we find that especially the Goldfeld–Quandt test rejects more often in the current study. In the test applied here, the variance of returns in the latter third of the sample is compare to that in the first third. As the institutional structure of the financial markets was much different in the late 1980's compared to that of the early 1970's, an increase in volatility is almost to be expected. Thus it is not really surprising that this test rejects more often than in comparable studies. The other tests applied here reject in about 25-30% of the time: this is a figure that is quite in line with the other studies. Thus there is no basis for asserting that the process generating returns to stocks on the Stockholm exchange during the 1970's and 1980's was qualitatively different than the processes on other stock exchanges.

Rather than concentrate on the heteroscedasticity of returns, some of the papers in the journals have asked whether the coefficients of the market model were stable in the sample that they studied. As this is the basic question asked here, a quick review of these papers is called for. The emphasis here is on tests of coefficient stability for specific models rather than the general question of beta stability which was addressed in Chapter 1. The tests usually consist of the estimation and comparison of one or more different models. Fabozzi & Francis (1978) estimate beta as a random coefficient — a constant with random disturbances. They use data from December 1965 to December 1971 and consider 700 stocks where data is available for all 73 periods. They test if the variance of the estimated deviations of the coefficients from their respective trend value — $(b_{it} - \beta_i)$ — is significantly different from zero. They find such significance in 15% of the cases (103 of the 700) when testing at the 10% level and in 8% (57) at the 5% level.

Sunder (1980) and Simonds *et al.* (1986) test whether the individual security's *beta* may be described as a random walk. Alexander *et al.* (1982) do similar tests for mutual funds. Sunder finds that he may reject the null hypothesis 77% of the time for the period 1926 to 1950 and for 39% for the period 1951 to 1975. Simonds *et al.* reject between 11% and 31% of the time depending on the period studied. Alexander *et al.* come to similar conclusions — their rejection rate varies is about 26% for the LaMotte–McWhorter test, 10% for the Sunder and 34% for either one or the other. DeJong & Collins (1985) compare the *RW*, the *RCF* and the *MRV* models but present only summary statistics — they claim no interest in finding the "correct" model.

Bos & Newbold (1984) find even greater evidence of random coefficients. Their data from the New York Stock Exchange encompassed the period from January 1970 to December 1979. They employed maximum likelihood methods to test various restrictions on the model

$$R_{it} = \alpha_i + \beta_{it} x_t + \varepsilon_{it} \qquad (2.33)$$
$$\beta_{it} - \bar{\beta}_i = \phi \left(\beta_{it-1} - \bar{\beta}_i \right) + a_t \qquad (2.34)$$

Testing the hypothesis that the variance of a_t is zero against the compound hypothesis that ϕ is zero and the variance of a_t is not, they find evidence of a random coefficient model in 58% of their stocks (270 of 464). They also find that allowing ϕ to be non-zero gives significant autoregressive coefficients for 16% of the stocks (75 of 464). They do not test for the random walk which obtains by setting $\phi = 1$. Collins *et al.* (1987) also estimate the random coefficient model. Using a sample size comparable to the present on, they reject the null hypothesis of constant coefficients against that of a random coefficient model in 28% of their 500 stocks. The corresponding percentage for their entire 500 month sample was 64%: the longer the period, the more instable the coefficients.

Using Australian data, Faff *et al.* (1992) also test the above model. They have a rejection rate of between 21% for 1978–1982 and of 29% for 1983–1987. In a companion study, Brooks *et.al.* (1992) also compare the *MRV* to the *RCF* model and find the evidence overwhelmingly in favor of the latter. Using the same time periods, they find, for 1978–1982, that the *RCF* model is preferred for 30 of the 34 cases where the null hypothesis of *beta* stability is rejected; for the latter period, the advantage of the *RCF* model is smaller — 59 of the 76 cases.

Using Finnish data, Knif (1989) estimates all three of the models (1.9)–(1.11). However, he refrains from selecting a "best" one. Instead he presents a large number of tables and allows the reader to draw his own conclusions. For example, if one sets a high preference for models where the updated state yields a return which lies close to the true observation, then the *RCF* model dominates: it is best by this criterion in 38 of the 39 cases with the odd one being the *MRV* model. If, however, one prefers good one–step ahead predictions, then his results are not as clear cut: the *MRV* and the *RCF* models perform about equally well. The poorest performance is returned by the *RW* model. Finally, if ones tastes run towards the Akaike Information Criterion, the well known *AIC*, then the constant coefficients model is best. I find this strange, but the results are very clear (see Knif (1989), p. 189).

Turning to the question of which model is "correct" for our data, we claim that *a priori* if either the LaMotte–McWhorter or the Sunder test is positive, then it seems reasonable to accept the random walk model. If, however, these two tests cannot reject the null hypothesis of stability and either the White or the Breusch–Pagan test do reject, then the test results suggest that the random coefficient model is the correct one. Thus, on the basis of the tests, we should find 15 *RW* models and 4 *RCF* ones. That is to say, 71% of the models are random walks and 3% random coefficients.[11]

2.4.3. WHY DIFFERENT TESTS YIELD DIFFERENT RESULTS

One of the more striking features of Tables 2.1 and 2.2 is the seemingly inconsistent results. If one excepts the Fluctuations test which explicitly asks whether regression coefficients are constant and the modified von Neumann test which along with the Box-Ljung test is really a test of general model adequacy, tests for heteroscedasticity, one feels, should give similar results.

However, differing test statistics have been constructed using differing alternative hypotheses. Thus it is not at all surprising that different tests give different results as they each ask different questions. To see this, let us rewrite the correct equation as follows:

$$y_t = \alpha + \beta r_t + \gamma z_t + \varepsilon_t \tag{2.35}$$

Here r_t and γ_t are the T dimensional vectors of the returns to the market portfolio and another variable, respectively. The estimated equation, (2.36) contains but a constant and a single regressor, r_t. Note that z_t is omitted

---

[11]Using the test results for the entire sample, we would expect about 35% of the models to be random walks and about 26% to be *RCF*. With the exception of Sunder's earlier period, these figures are well in line with the findings of others.

from the equation. α, β and γ are the respective regression coefficients and ε_t is the error term. What is estimated is therefore

$$y_t = \tilde{\alpha} + \tilde{\beta} r_t + \mu_t \qquad (2.36)$$

where the error term is now

$$\mu_t = \gamma z_t + \varepsilon_t \qquad (2.37)$$

Assume now that z_t can be approximated as a function of r_t as follows:

$$z_t = \lambda_0 + \lambda_1 r_t + \lambda_2 r_2^2 + v_t \qquad (2.38)$$

Again, terms with a higher order than two may be included, but a second degree equation will suffice. Substituting (2.38) into (2.35) we find that

$$\tilde{\beta} = \beta\gamma\lambda_1 + \beta\gamma\lambda_2 r_t \qquad (2.39)$$

Note that the estimated regression coefficient has become time–varying: I pointed out earlier that non constant regression coefficients could yield positive tests for heteroscedasticity. White's test for heteroscedasticity assumes a relationship such as (2.38). The Breusch–Pagan test omits the linear terms in the independent variable and thus the difference between these two tests becomes apparent. Note that even the tests for random walk models fit into this pattern. Thus (2.24) may be thought of as a variant of (2.38) with all the λ–terms equal to one and the r_t–terms being multiplied by a random number instead of being raised to differing powers.

The *ARCH* test offers yet another specification for the expected value of the squared residuals in the *OLS* estimation such as (2.25). The Goldfeld–Quandt test as applied here assumes that this expected value is proportional — or inversely so — to time. Indeed, that this test is significant in only a little more than 50% of the cases is rather surprising even if other studies have shown a lower percentage. The Swedish stock market underwent a drastic change when the credit market was deregulated: volume increased dramatically as did prices. One would thus also expect the volatility of the stock returns to have increased over the period.

2.5. The cumulative and moving sum of squares tests

The results in this section are rather unusual in that the literature shows almost no applications of either the cumulative or the moving sum of squares tests to financial data. One exception is, however, Knif (1989) who tests for parameter stability using both the cumulative sum of squares and the moving sum of squares techniques. Employing data on 39 stocks from the

Helsinki Exchange for the period January 1970 to December 1985, he finds that almost all test positively for parameter variation.

There is of course a problem of interpreting the results. The basic model tested here is (2.33)–(2.34) with $\phi = 1$. "non constant regression coefficients" can mean that the variance of a_t is non-zero. It may also mean more or less discrete shifts in the parameters. Such shifts are excluded *a priori*. Coefficients may well follow some stochastic process as in (2.34) while the cumulative and moving sum of squares plots give no evidence of shifts in parameters. However, if *beta* becomes more volatile, the variance of a_t will increase. This in turn will induce a peak above the confidence level line in the moving sum of squares diagram. It will also tend to make the plot of the cumulative sum of squares flatter before the shift and steeper after it.

My results are similar to Knif's. He found that at the 10% level, 36 of his 39 stocks showed evidence of structural change when tested using the cumulative sum of squared residuals and the 35 of the 39 when using the moving sum of squared residuals test.[12] He does not indicate whether the same stocks tested negatively for both tests. I find that only 5 of the 57 stocks show no evidence at the 10% level of structural change; for the 21 stocks presented here, all but Pharos reject the null of no structural change. For 16 of these stocks, both tests indicate some such change; for 4 only the *cusumsq* test is positive; and in no case did only the moving sum of squares test suggest structural change. I might note that the window chosen for the moving sum of squares test is 12. Also there are two cases — Garphyttan and Munksjö — where the *cusumsq* line crosses both the upper and the lower confidence bound. Table 2.3 present the results in a compact form. There the maximum absolute deviation of the cumulative sum of squared residuals from the line from $1/(T - k - 1)$ to 1 along with the probability that this value obtains in a two sided test is tabulated.[13] The table also contains the maximum value for the moving sum of squared residuals and its corresponding 10% confidence level. This confidence level is constructed using an inequality that works to satisfaction when the probability of a given observation is small, but of little use when the probability is as large. Here "large" should be taken to mean 20% or higher. Thus only the upper

---

[12] His window here was 10. With a window of 15 or of 20, he found that all stocks showed evidence of structural change; with a window of 5 only 12 of 39 tested positively. Why he used multiples of 5 to test monthly data is not explained in his book: multiples of 6 would seem more logical.

[13] In an appendix to this chapter I present a method to expand the table for the *cusumsq* confidence levels given by Durbin and widely reproduced. I also present a table to complementing Durbin's and a formula for approximating other values to an accuracy of at least 0.01 percent. Such a figure is more than sufficient for the purposes of graphical analysis. For summary statistics, the maximum deviation of *cusumsq* from the line between $1/(T - k + 1)$ and 1 as well as its calculated probability may be used.

TABLE 2.3. **Tests using the** *cusumsq* **and** *mosumsq*. An entry of "0.000" indicates that the *p-value* was less than 0.0005.

Stock	T	max cs^2	p-level	max ms^2	p-level
Aga	252	0.131	0.020	2.436	3.103
Alfa Laval	247	0.188	0.000	3.686	3.101
Asea	252	0.212	0.000	5.039	3.103
BGB	104	0.266	0.001	2.678	3.083
Custos	251	0.265	0.000	3.419	3.102
Ericsson	252	0.290	0.000	4.819	3.103
Esab	252	0.132	0.011	4.869	3.103
Export	249	0.162	0.002	5.601	3.102
Gambro	104	0.203	0.017	4.849	3.083
Garphyttan	252	0.243	0.000	8.318	3.103
Industrivärden	251	0.165	0.001	3.210	3.102
Investor	251	0.243	0.000	4.826	3.102
Munksjö	238	0.239	0.000	8.104	3.097
Pharmacia	235	0.246	0.000	5.437	3.096
Pharos	131	0.114	0.277	2.181	3.072
Protorp	131	0.158	0.052	2.982	3.072
Providentia	251	0.199	0.000	4.827	3.102
SKF	252	0.233	0.000	5.884	3.103
Trelleborg	251	0.180	0.000	5.133	3.102
Volvo	252	0.114	0.059	2.371	3.103
Åkermans	239	0.285	0.000	9.079	3.098

confidence level is calculated.

These two tests may be complemented by running the calculations backwards. This is useful when the parameter shifts are suspected to have occurred near the end of the sample. However, I find that these tests run backwards almost always point in the same direction as when the cumulative or moving sums are calculated in the usual manner.[14]

While the maximum values of these two statistics give a parametric test,

[14]For the 21 presented here, there was no noticeable difference between running the sample forward or backwards. However, using the entire sample, I found that the *cusumsq* for Korsnäs and Perstorp gave conflicting results: the maximum value of the statistic for the former is significant at the 10% level when the calculations go backwards in time but not when they go forward; the opposite holds for Perstorp — the forward but not the backwards calculations reject the null hypothesis. The lone exception using the *mosumsq* test is Siab: the forward calculations give a significant test statistic.

there is still much information to be had from graphs. I present therefore these graphs on the following pages with the *cusumsq* diagram being the upper most of the pair and the *mosumsq* the lower. The diagrams show they type of plots that are derived from the data. Structural change is indicated by abrupt changes in the slope of the *cusumsq* line or sharp, high peaks in the *mosumsq* line. However, placing a date on the shift is really more an art than a science. Examining the diagrams on the following pages, we find that only one of the stocks shows no evidence of structural change — Pharos — while Aga and Volvo show marginal evidence (the *cusumsq* line crosses a confidence line but there is no well defined change in slope nor does the *mosumsq* line peak). Perhaps BGB and Protorp, whose *mosumsq*'s lack a peak while their *cusumsq*'s are rather smooth, should also be included in the group that shows but marginal evidence of structural change. Trelleborg with its sharp peak in the *mosumsq* and a hop in the *cusumsq* is a typical example of a stock with a structural break. The rest of the twenty show definite breaks. Asea, Ericsson, Munksjö, and Pharmacia show definite signs of change in the early 1980's while Gambro's break comes a bit later on in the decade. Especially the peaks around 1983 are observed in many of the series — a rough count shows about 10 of the 21 show some peak in their *mosumsq* here. There is also another top in the later part of the period — around 1990: 13 of the series give evidence of an increase in volatility here but the data do not allow us to identify a structural change at the end of the period. Surprisingly few show any signs of structural change in 1987, but as the entire market fell, most of the stocks seemed to fall with it and thus the residual in the market model equation does not show an unusual increase in variance. Note, finally, that Investor, SKF and Åkermans all show structural breaks at the end of the period.

2.6. Some preliminary conclusions

Before drawing any sweeping conclusions about structural stability or variable regression coefficients, I will present one more diagnostic tool, flexible least squares. This method involves iterating the Kalman filter and I will devote an entire chapter to the method.

None the less, some generalizations may be made here. Pretesting gives one a general feeling for the data at hand and is helpful when formulating a model. However, as different tests are based on different assumptions and designed to test different aspects of the data, one should not be surprised that different tests give different answers.

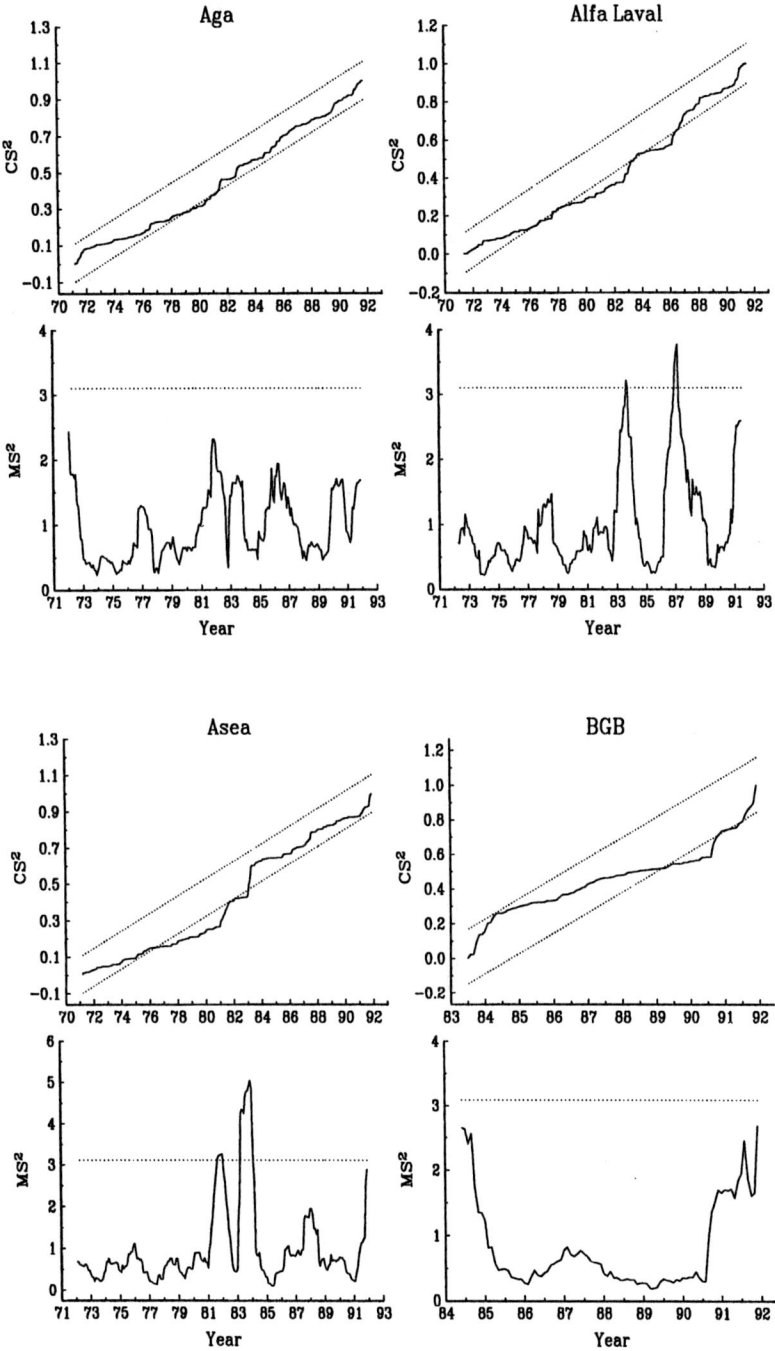

Figure 2.1. **Cusumsq and Mosumsq estimates.** The top diagram of each pair is the *cusumsq* estimate.

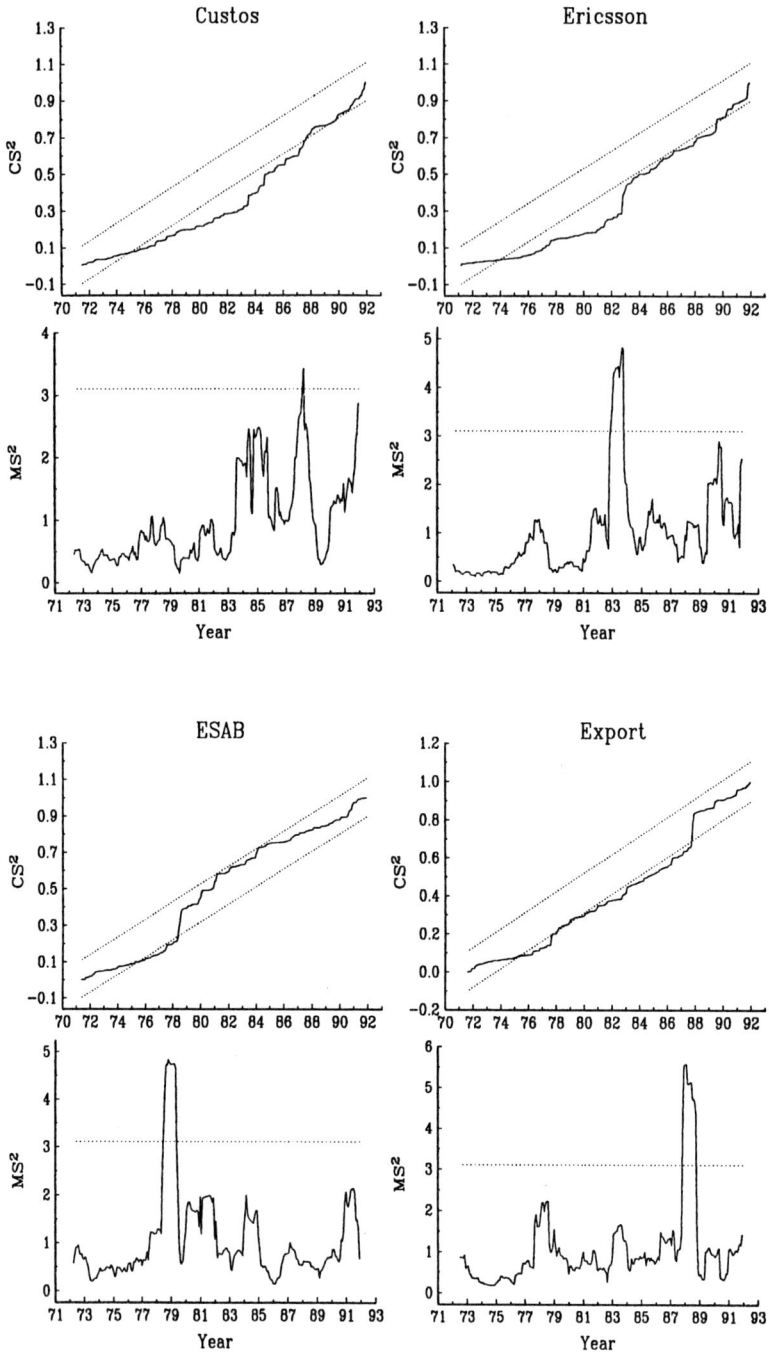

Figure 2.2. **Cusumsq and Mosumsq estimates.** The top diagram of each pair is the *cusumsq* estimate.

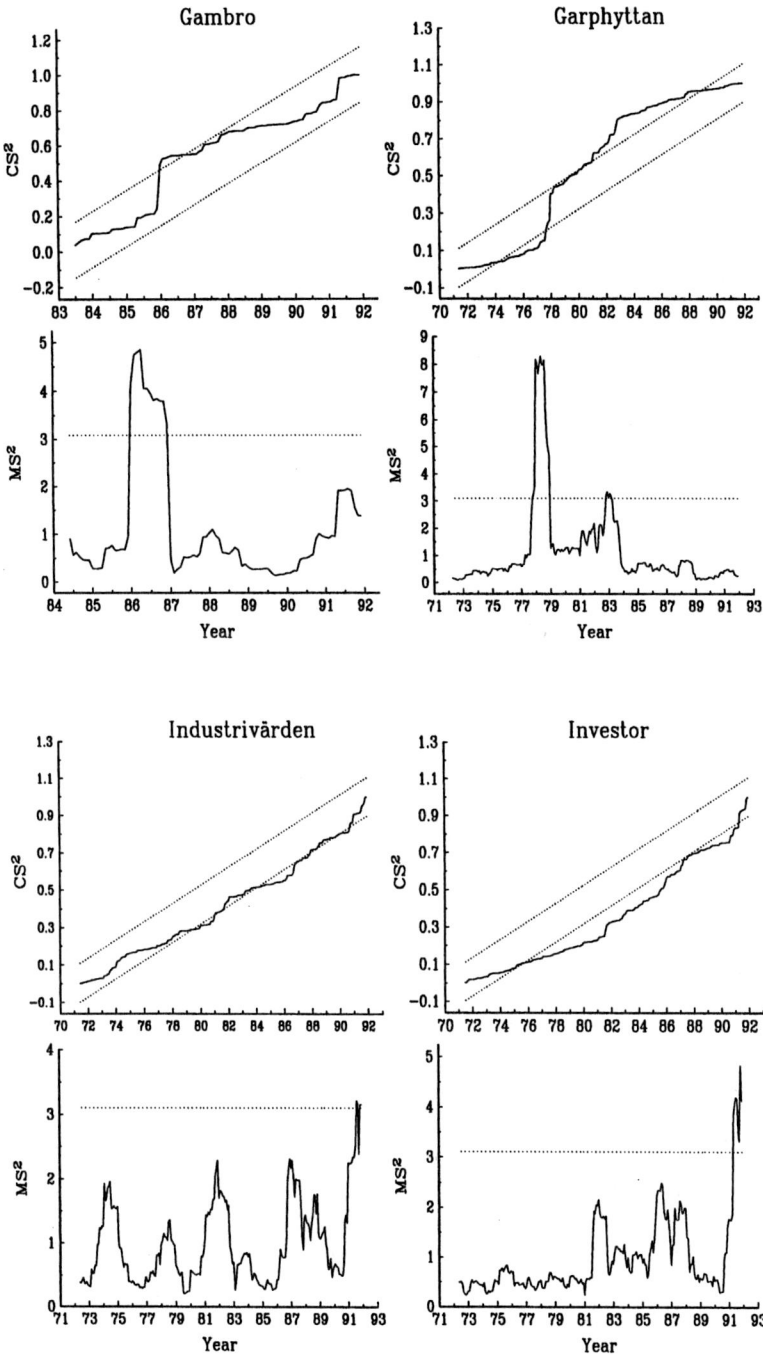

Figure 2.3. **Cusumsq and Mosumsq estimates.** The top diagram of each pair is the *cusumsq* estimate.

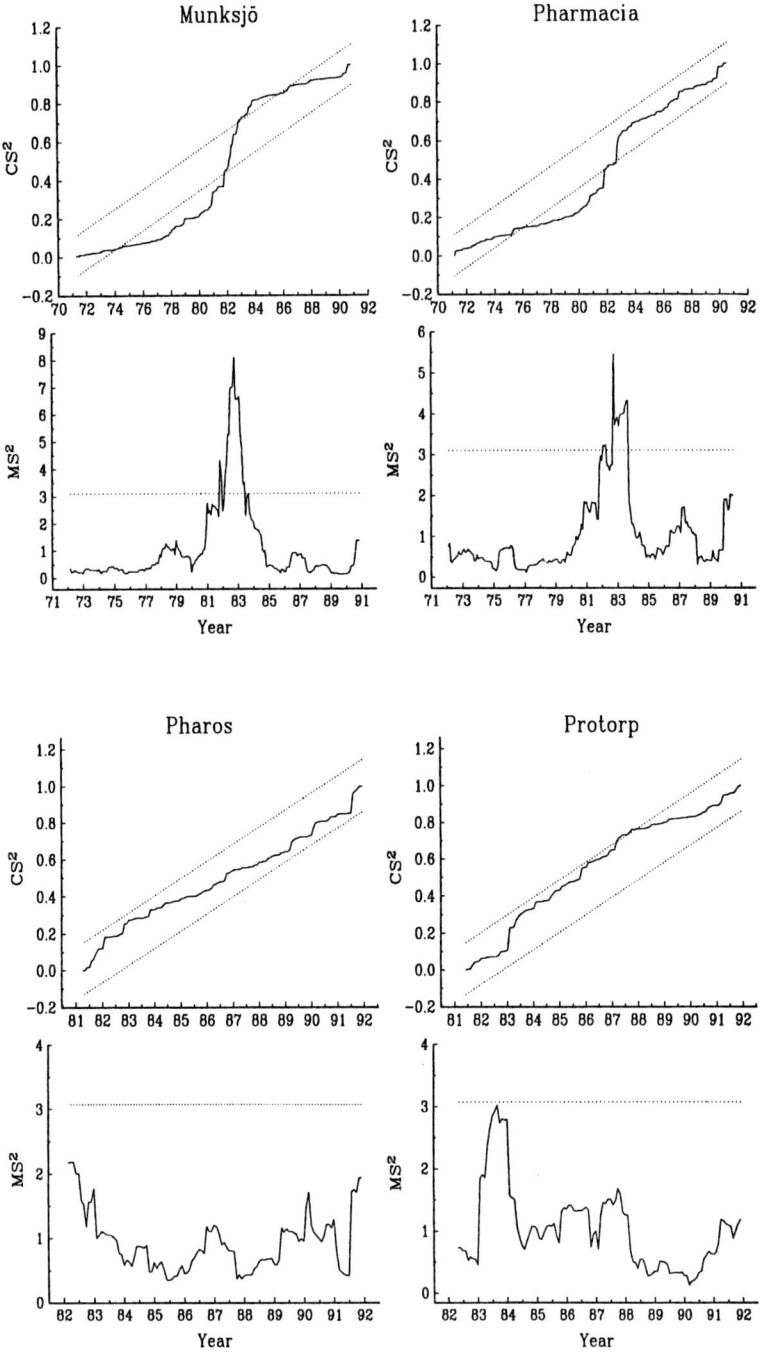

Figure 2.4. **Cusumsq and Mosumsq estimates.** The top diagram of each pair is the *cusumsq* estimate.

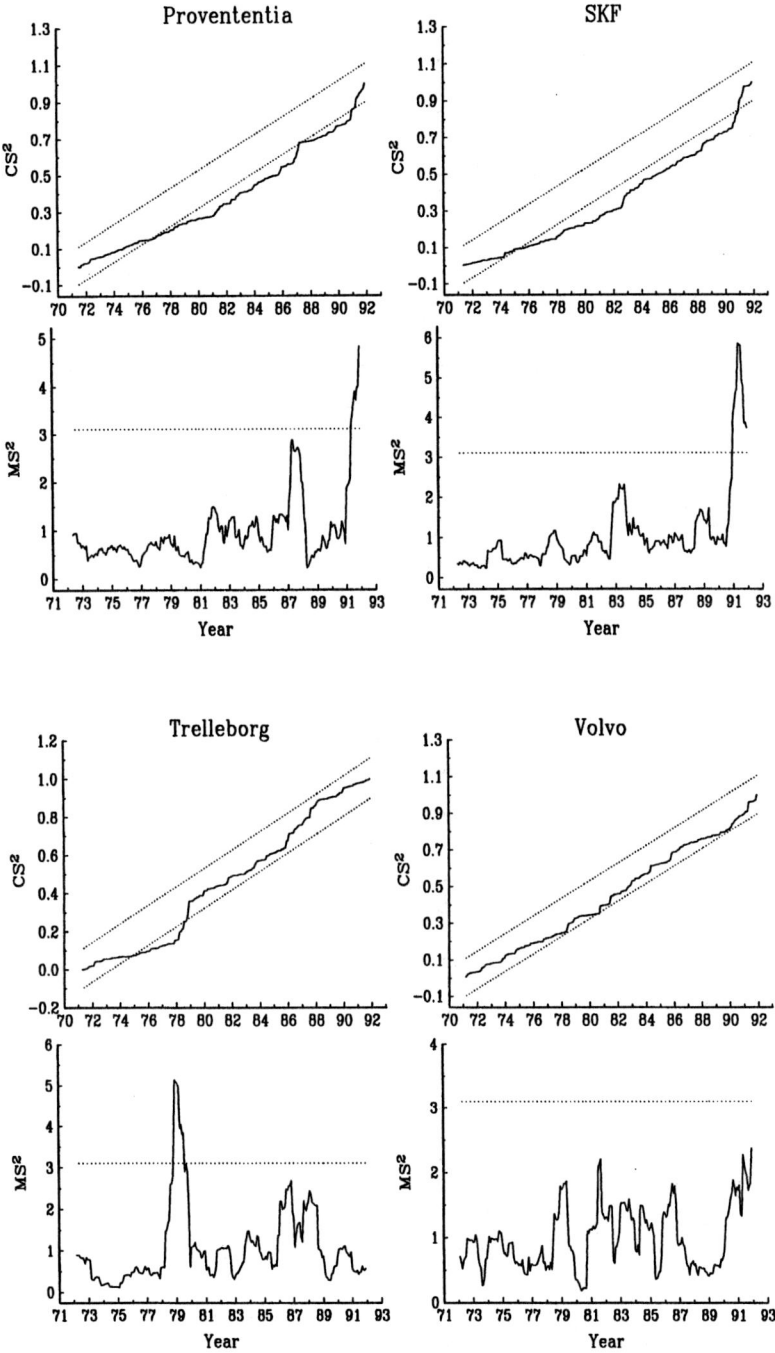

Figure 2.5. **Cusumsq and Mosumsq estimates.** The top diagram of each pair is the *cusumsq* estimate.

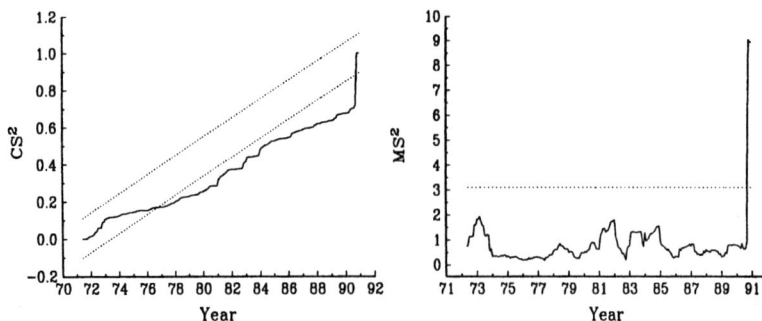

Figure 2.6. **Cusumsq and Mosumsq estimates for Åkermans.** The left hand diagram is the *cusumsq* estimate. Note that if the final six months are removed from the sample, then there remains no evidence of a structural break.

Secondly, one should choose a rather loose critical level — narrow intervals such as the 5% or perhaps even the 10% confidence bounds are preferred to wider ones as rejecting a true null hypothesis of stability is a much lesser error than failing to reject a false null. After all, if everything is stable, then the estimates based on instability will be insignificant. The course of action taken here is to assume instability when one of the tests rejects the null.

Finally, the *cusumsq* and the *mosumsq* statistics should be plotted: there may be information in the diagram that is not obvious from the summary statistic. For example, when a structural break occurs at the end of the period, there is not much than can be done other than to observe the change. Especially, Åkermans but also Investor and SKF fall into this category. Removing the last six months of the sample for Åkermans reverses our conclusions on structural stability. However, both Investor and SKF, once the last year is removed, show structural breaks in the early 1980's. Without the diagrams such inferences cannot be made.

2.7. Appendix: The cusumsq statistic

Since its introduction by Brown *et.al.* (1975) the *cusumsq* statistic has become a popular test of parameter stability in regression models. The usefulness of the test depends not only on its relative simplicity, but also on the emphasis the original authors paid to a graphical presentation. In favorable circumstances this can lead to very plausible interpretations.

The basic idea of the test is to compare recursive estimates of the residual variance, which is technically achieved by plotting the ratios between partial and total residual sums of squares as in (2.19). Confidence bounds for this plot can be found using critical values tabulated by Durbin (1969) in connection with another test. The advantage of this procedure compared with, *e.g.*, recursive Chow tests, is that the joint significance level is correctly given. Sharp turning points in the *cusumsq* plot also indicate likely time points for structural change.

Two practical problems connected with the use of the *cusumsq* test have been the non-availability of an algorithm for calculating P–values, and sufficiently accurate critical values in medium and large sized samples such as are typically encountered in studies of financial data. These deficiencies have been corrected in a recent paper by Edgerton & Wells (1994). This appendix will summarize the main results of their paper, including an expanded table and the formula, but I refer the reader to the paper itself for greater detail.

As in section 2.2.2, consider T observations from a regression of a dependent variable y on a set of k independent variables. Letting z_t be the row vector of regressors at time point t, and $Z'_t = (z'_1, \ldots, z'_t)$ be the matrix of regressors up to time point t, then the $T - k$ recursive residuals v_t are defined as

$$v_t = \frac{y_t - z'_t b_{t-1}}{\sqrt{1 + z_t (Z'_t Z_t)^{-1} z'_t}}, \qquad t = k+1, \ldots, T, \qquad (2A.1)$$

where b_t is the least squares parameter estimate based on t observations. If all the usual regression assumptions are fulfilled (the null hypothesis), then the recursive residuals are independently distributed with the same distribution as the true stochastic errors. The recursive residuals are usually calculated from (2.17) using the recursions of equations (2.13)-(2.16). The *cusumsq* quantities are defined by equation (2.17) which is repeated here:

$$s_t = \left(\sum_{j=k+1}^{t} v_j^2 \right) / \left(\sum_{j=k+1}^{T} v_j^2 \right), \qquad t = k+1, \ldots, T. \qquad (2A.2)$$

Consider, now, only the *even* time periods with the *cusumsq* statistics $s_{k+1}, s_{k+4}, \ldots, s_{k+2n}$ constituting $n_v = [\frac{1}{2}(T - k - 1)]$. Note that $[r]$ is the modulus of r. Define

$$v^+ = \max_{t=1,\ldots,m-1} \left(S_{k+2t} - \frac{t}{m} \right) \qquad (2A.3)$$

and

$$v^- = \max_{t=1,\ldots,m-1} \left(\frac{t}{m} - S_{k+2t} \right) \qquad (2A.4)$$

where $m = \frac{1}{2}(T - k) - 1$. These two statistics follow the distribution given in Durbin (1969) with a tail probability of

$$P\left(v^+ > v_0\right) = F\left((n_v + 1) v_0, n_v\right) \qquad (2A.5)$$

where the function $F(a, n)$ is given by

$$F(a, n) = \frac{a+1}{(n+1)^n} \sum_{[a]+1}^{n} \binom{n}{j} (j-a)^j (n+1+a-j)^{n-1-j} \qquad (2A.6)$$

An analogous result holds for v^-. Critical values are found by solving

$$F\left((n+1) c_{n\alpha}, n\right) = \alpha \qquad (2A.7)$$

Durbin shows that the statistic $v = \max(v^+, v^-)$ has a tail probability that may be approximated by twice that given in (2A.7) relative error of less than 0.1

$$F\left((n+1) c_{n\frac{\alpha}{2}}, n\right) \approx \alpha \qquad (2A.8)$$

Using v^+ implies discarding half of the *cusumsq* quantities. However, it follows from the argument in Brown *et.al.* that the statistics based on the odd time periods will have the same distribution as above. The usual *cusumsq* test as suggested by Brown *et.al.* is to use the maximum absolute deviation from the reference line as the *cusumsq* statistic:

$$\xi = \max_{t=1,\dots,T-k} \left| s_{k+t} - \frac{t}{T-k} \right| \qquad (2A.9)$$

The number of degrees of freedom will be $n_\xi = \frac{1}{2}(T-k)-1$. The advantage of this approach is that all observations are used. The disadvantages are partly practical — n_ξ may not be an integer — and partly theoretical — the distribution of ξ is assumed to be equal to the distribution of v. This latter point is probably of little importance in large samples, but may be a problem in small ones.

n_ξ is not an integer when $T-k$ is odd. Defining $n_\omega = [\frac{1}{2}(T-k)] - 1$, the P-values for ξ may be calculated as

$$P_\xi = \frac{1}{2}F\left((n_v + 1) x, n_v\right) + \frac{1}{2}F\left((n_\omega + 1) x, n_\omega\right) \qquad (2A.10)$$

where x is the observation of ξ.

Durbin's tables list the critical values for n_ξ between 1 and 60 as well as for even values up to 100. There is no real problem in expanding Durbin's table to larger degrees of freedom: the problems are numerical rather than theoretical. The factorial term becomes exceeding large as do the second of the two exponential terms as n_ξ increases past 100. To avoid numerical underflow or overflow, one must rewrite (2A.6) so that all exponential terms are approximately unity raised to a power. The factorial terms must also be kept near one which is accomplished by calculating the formula recursively. By defining $\Theta = [a] + 1$, (2A.6) has been rewritten as

$$S_m = (a+1) A_m \sum_{j=\Theta}^{m-1} b_{m,j} \tag{2A.11}$$

where

$$A_m = \frac{(m-1)!}{m^{m-1}} \cdot \frac{e^m}{\sqrt{m}} \tag{2A.12}$$

$$b_{m,j} = e^{-m}\sqrt{m} \cdot \frac{(j-a)^j}{j!} \cdot \frac{(m-a+j)^{m-2-j}}{(m-1-j)!} \tag{2A.13}$$

The constant A_m approaches in the limit $\sqrt{2\pi}$ as may be seen using Sterling's formula:

$$m! = \sqrt{2\pi} \cdot m^{m+\frac{1}{2}} \cdot e^{-m}$$

As m grows without bound, it is apparent that the (2A.12) approaches $\sqrt{2\pi}$.

A recursive formula for calculating $b_{m,j}$ obtains from (2A.13):

$$
\begin{aligned}
b_{m,j+1} = & \frac{j+1-a}{j+1} \cdot \frac{m-1-j}{m+a-j-1} \cdot \left(1 + \frac{1}{j-a}\right)^j \\
& \cdot \left(1 - \frac{1}{m+a-j}\right)^{m-2-j} \cdot b_{m,j}
\end{aligned}
\tag{2A.14}
$$

given the initial value

$$b_{m,\Theta} = \frac{(\Theta-a)^\Theta}{\Theta!} \cdot \frac{(m+a-\Theta)^{m-2-\Theta}}{(m-1-\Theta)!} \cdot e^{-m} \cdot \sqrt{m} \tag{2A.15}$$

When m is large the term e^{-m} in (2A.15) will cause an underflow. This problem may be avoided for moderately large m by taking the natural logarithm of both sides and adding instead of multiplying. However, as m gets really large, even $\exp(\ln b_{m,\Theta})$ may cause an underflow. This last problem is avoided by checking the size of $\ln b_{m,\Theta}$ and, if it is too small to be handled by the machine, increasing Θ by one and updating the formula using (2A.15). Eventually a suitable starting value for Θ will be reached and recursions in (2A.14) will begin using this initial value for $b_{m,\Theta}$. I should add, that, on my machine, "too small" means a number around $\exp(-307)$.

Iterating (2A.6) for a calculated *cusumsq* statistic gives the probability of obtaining that statistic given the null hypothesis of structural stability.

By setting (2A.6) to a desired probability level, one defines an equation in a which may be solved by standard quasi-Newton methods. This solution is then the table entry for m degrees of freedom. There is one problem that should be addressed. If m, is an even number, then calculating the probability is straight forward. However, if it is an odd number, then division by two produces a non integer number. One avoids this problem using the formula in the same way one avoids it when using the table: by linear interpolation. Calculate the probability for $m/2+0.5$ and for $m/2-0.5$ and use the average as the probability. Tables 2.5 and 2.6 extend Durbin's table to cover the odd degrees of freedom form 61 to 99. It continues at steps of 10 up to 200 degrees of freedom; and then with steps of 100 from 200 to 1000 and finally by steps of 1000 from 1000 to 10000 degrees of freedom.

Even if the recursions above are sufficient for finding the tail probability of the calculated statistic, graphical analysis requires confidence intervals. This of course may be obtained from a table or calculated from (2A.6) if the sample at hand does not correspond to a tabulated one. There is, however, another possibility.

As the number of observations grow we would hope to find an adequate asymptotic approximation, and thus avoid the need to consult tables. It can be shown that the asymptotic distribution of ξ is the same as that of the Kolmogorov–Smirnov (KS) statistic, and that an asymptotic approximation to the critical values is thus given by

$$c_{n\alpha}^0 = \sqrt{\frac{-\ln \alpha}{2n}} \tag{2A.16}$$

This approximation is fairly inaccurate, however, yielding a relative error of over 5% already when $n = 100$. Following a suggestion of Miller (1956), Edgerton & Wells propose the following extension to (2A.16):

$$c_{n\alpha}^0 = \frac{a_{1\alpha}}{n^{1/2}} + \frac{a_{2\alpha}}{n} + \frac{a_{3\alpha}}{n^{3/2}} \tag{2A.17}$$

where $a_{1\alpha} = \sqrt{-0.5 \ln \alpha}$ is obtained from the asymptotic value of the KS-statistic. The other two constants were estimated by least squares, using the exact values for $n = 100$ to 10000 as the dependent variable, and the values in Table 2.4 were obtained. This approximation is extremely accurate, the coefficient of determination being in all cases larger than 0.999999. The percentage error for the approximation compared to the true value is never larger than 0.0049100). Indeed, at $m = 30$ which is of course out of sample, the error is about 0.2%. The relative error is less than two percent with n as small as 10. For n greater than 100 the absolute error is only evident in the sixth decimal place. When plotting, errors of this size are not noticeable.

TABLE 2.4. Coefficients in the *cusumsq* approximation.

	$\alpha = 0.10$	$\alpha = 0.05$	$\alpha = 0.025$	$\alpha = 0.01$	$\alpha = 0.005$
a_0	1.0729830	1.2238734	1.3581015	1.5174271	1.6276236
a_1	-0.6698868	-0.6700069	-0.6701218	-0.6702672	-0.6703724
a_2	-0.5816458	-0.7351697	-0.8858694	-1.0847745	-1.2365861

Thus one could use the approximation even if table values are available. As the formula works for non-integer as well as integer values, the need to interpolate for half degrees of freedom thereby disappears.

The tables on the following pages are compliments to Durbin's table. They contain the critical values for the odd integers between 61 and 99 which were excluded from the original table. Financial data sets often contain more than the 200 or so data points covered by the original table. Thus the tables presented here also contain the critical values for as many as 10000 degrees of freedom.[15]

---

[15]There is also a GAUSS routine that calculates the exact critical value for any given number of degrees of freedom. See page 151 for details on downloading this routine from our *ftp* server.

TABLE 2.5. **Critical values for the** *cusumsq* **statistic, n = 61 - 200.** Enter the table at the value
$n = 0.5 * (T - k) - 1$. For a two sided test, double
the confidence level. *Source:* Edgerton & Wells (1994),
p. 360; reprinted by permission of the *Oxford Bulletin
of Economics and Statistics.*

n	$\alpha=0.10$	$\alpha=0.05$	$\alpha=0.025$	$\alpha=0.01$	$\alpha=0.005$
61	0.12522	0.14422	0.16109	0.18107	0.19486
63	0.12342	0.14213	0.15874	0.17841	0.19199
65	0.12170	0.14013	0.15649	0.17587	0.18925
67	0.12006	0.13821	0.15433	0.17344	0.18662
69	0.11848	0.13637	0.15227	0.17110	0.18410
71	0.11696	0.13461	0.15028	0.16886	0.18168
73	0.11550	0.13291	0.14838	0.16671	0.17936
75	0.11409	0.13128	0.14654	0.16463	0.17712
77	0.11273	0.12970	0.14478	0.16264	0.17497
79	0.11143	0.12819	0.14307	0.16071	0.17289
81	0.11017	0.12672	0.14143	0.15886	0.17089
83	0.10895	0.12531	0.13984	0.15706	0.16896
85	0.10777	0.12394	0.13831	0.15533	0.16709
87	0.10663	0.12262	0.13682	0.15366	0.16528
89	0.10553	0.12134	0.13538	0.15204	0.16353
91	0.10446	0.12010	0.13399	0.15046	0.16184
93	0.10342	0.11889	0.13264	0.14894	0.16020
95	0.10241	0.11773	0.13134	0.14747	0.15861
97	0.10144	0.11660	0.13007	0.14603	0.15706
99	0.10049	0.11550	0.12883	0.14464	0.15556
100	0.10002	0.11496	0.12823	0.14396	0.15483
110	0.09571	0.10997	0.12263	0.13765	0.14803
120	0.09193	0.10558	0.11772	0.13211	0.14206
130	0.08856	0.10169	0.11336	0.12720	0.13676
140	0.08555	0.09821	0.10946	0.12280	0.13202
150	0.08283	0.09506	0.10594	0.11884	0.12775
160	0.08035	0.09220	0.10274	0.11524	0.12387
170	0.07809	0.08959	0.09982	0.11195	0.12033
180	0.07601	0.08719	0.09714	0.10893	0.11708
190	0.07409	0.08498	0.09466	0.10614	0.11408
200	0.07231	0.08293	0.09237	0.10356	0.11130

TABLE 2.6. **Critical values for the** *cusumsq* **statistic, n = 300 - 10000.** Enter the table at the value $n = 0.5 * (T - k) - 1$. For a two sided test, double the confidence level. *Source*: Edgerton & Wells (1994), p. 360; reprinted by permission of the *Oxford Bulletin of Economics and Statistics*.

n	$\alpha=0.10$	$\alpha=0.05$	$\alpha=0.025$	$\alpha=0.01$	$\alpha=0.005$
300	0.05960	0.06828	0.07601	0.08517	0.09150
400	0.05190	0.05943	0.06612	0.07406	0.07955
500	0.04659	0.05333	0.05932	0.06642	0.07134
600	0.04265	0.04880	0.05427	0.06076	0.06525
700	0.03957	0.04526	0.05033	0.05634	0.06049
800	0.03707	0.04240	0.04714	0.05276	0.05665
900	0.03500	0.04002	0.04449	0.04980	0.05346
1000	0.03324	0.03801	0.04225	0.04728	0.05076
2000	0.02365	0.02702	0.03002	0.03358	0.03605
3000	0.01936	0.02212	0.02457	0.02747	0.02949
4000	0.01680	0.01918	0.02130	0.02382	0.02556
5000	0.01504	0.01717	0.01907	0.02132	0.02288
6000	0.01374	0.01569	0.01742	0.01948	0.02090
7000	0.01273	0.01453	0.01614	0.01804	0.01936
8000	0.01191	0.01360	0.01510	0.01688	0.01811
9000	0.01124	0.01283	0.01424	0.01592	0.01708
10000	0.01066	0.01217	0.01351	0.01511	0.01621

Chapter 3

Flexible Least Squares

3.1. Introduction

I have argued above that in a changing environment, the assumption of constant coefficients is unrealistic. However, nothing is impossible: certain relationships may just happen to be characterized by a stable regression e-quation. Therefore one must examine the data at hand before one accepting or rejecting stable coefficients as a working hypothesis. In what follows here I will subject the data to a rather *a theoretical* test for stability. The method used is similar in spirit to that which I will present in a later chapter dealing with maximum likelihood estimation. While the mathematics may be similar, the reasoning behind the calculations is not.

The approach outlined below is called *flexible least squares* or *flexible least cost* by Kalaba & Tesfatsion (1989) who suggest the method. The basic idea is that a model — and in particular an economic or social science model — is never exact. There is always a residual that must be explained by omitted variables, errors in measurement connected with the variables in the model or perhaps even by an incorrect mathematical specification of the model. For example as *CAPM* model is really an equilibrium relationship, disequilibrium situations will appear as errors in the model. One way to model this residual is to give it a probability distribution and use estimation methods to find acceptable parameters. This will be done in a later chapter. What Kalaba & Tesfatsion assume is that the researcher may hesitate in assigning probability densities to the residual terms. Anderson & Moore (1979) suggest a number of ways to accommodate this residual in the model. They suggest that observations be weighted so that newer ones receive more emphasis or that the noise parameters be increased so that the residual will become a larger part of the problem.

Here one should note while that control engineers have physical models to follow, economists must hypothesize a structure before estimation. Thus noise parameters in economic models are typically unknown factors and must be estimated. The basic approach is to assume that there are two sources of error in a model. First is what might be called a *measurement error*. This will exist because the specified relationship is not exact. The "true" relationship may be nonlinear, for example, but the specification calls for a linear model. There are, however, other factors that will affect this error term. A variable may have been omitted from the equation and those that are included may be observed with some error. The square of the deviation of the observation from the value predicted by the model will be the numerical value attached to the measurement error for each point in time.

The second error in the model is a *dynamic error*. This error occurs as the regression coefficients have been specified as constants but may in fact be variable. To have operational meaning, this error requires there to be a dynamic model for the coefficients. We will assume that these develop as random walks, but this is not the only specification possible. Indeed, as the dynamic equation will be a system of first order difference equations, it is but a simple matter to expand the coefficient vector to accommodate higher order systems. Of course, the entire model must be modified but such changes are minor and will not be considered here.

The next section specifies the problem more concretely. The model and the function to be optimized will be present along with the solution. The third section will present a number of examples of this solution. The final section will draw tentative conclusions.

3.2. The least cost solution

The model studied here is a simple one. The basic equation at time t is simply a regression equation relating the dependent variable y_t to a row vector of n independent variables z_t. The regression coefficients b_t are thus an n–dimensional column vector.

$$y_t = z_t b_t + \varepsilon_t \qquad (3.1)$$

In the illustrations that follow in the next section, z_t will be T by 2 with the first column consisting of ones. The coefficient vector at time t will be called the *state vector*. It is assumed to follow the following dynamic path:

$$b_{t+1} = \Phi b_t + \xi_{t+1} \qquad (3.2)$$

The matrix Φ is the transition matrix that translates the state at time $t - 1$ to the state at t. It is assumed to be an identity matrix as the states develop as random walks.

3.2.1. THE COST FUNCTION

Given the above equations, the following two costs may be defined. First we have a *measurement* or *system cost* which measures how well the model explains the observations:

$$r_m^2 = \sum_{t=1}^{T} (y_t - z_t b_t)^2 \tag{3.3}$$

This cost is simply the sum of the measurement error at each point in time. A more familiar name would be the sum of the squared residuals. The noisier the system is — that is, the larger the residual — the larger the measurement cost. Next consider the weighted deviation in the states from one period to the next:

$$r_\xi^2 = \sum_{t=1}^{T} (b_{t+1} - \Phi b_t)' W (b_{t+1} - \Phi b_t) \tag{3.4}$$

This cost is the weighted sum of the squares of the dynamic error at each point in time. This *dynamic cost* serves to keep the system "honest". That is, if we allow the states to vary greatly from period to period, then we can reduce the measurement error in each period to zero. To see this, in equation (3.1) set $\varepsilon_t = 0$:

$$y_t = z_t \cdot b_t \tag{3.5}$$

As z_t is not square — it is n by 1 — one cannot solve for b_t by inverting the matrix. Indeed, there is an infinite number of solutions to the equation (3.5). One of these obtains by using the generalized or *pseudo* inverse:

$$b_t = (z_t z_t')^{-1} z_t' y_t \tag{3.6}$$

The above expression is of course similar to the least squares solution to the problem. But unlike a regression equation, this one holds exactly as z_t has as many columns as b_t has rows. While this trick removes measurement error, the cost is a state vector with extreme volatility.

Nor is there any real problem involved in removing the entire dynamic cost from the system. By setting ξ_t to the null vector the states will not vary at all. If we imagine that b_t in (3.6) is a n–dimensional vector of constants, that z_t is a T by n matrix and that y_t is a T–dimensional vector, then this equation is the regression solution to the problem and the measurement error becomes the sum of the squared residuals for the regression.

Consider, now, the following cost functional:

$$C(b_t, \lambda) = \sum_{t=1}^{T} (y_t - z_t b_t)^2 + \lambda \sum_{t=1}^{T} (b_{t+1} - \Phi b_t)' W (b_{t+1} - \Phi b_t) \qquad (3.7)$$

The total modeling cost is dependent upon both the measurement cost and the dynamic cost multiplied by a constant λ. Equation (3.7) includes a positive definite weighting matrix W which is set to the identity matrix in Section 3.3 but included here for completeness.[1]

Given λ and W, (3.7) may be minimized with respect to the trajectories of the states. The minimum thus obtained may be divided into the two basic components, the measurement cost and the dynamic cost. These two could be plotted in a diagram with the respective costs on the axes. The point thus obtained would show the lowest possible cost obtainable given the two weights. Repeating the minimization with for values of λ between zero and some large number, one may trace out a frontier showing the relationship between the dynamic and measurement costs. The slope of this frontier represents the marginal trade-off between these two costs. As λ approaches zero, the frontier approaches the horizontal axis and measurement cost also approaches zero. If the slope at — or close to — the intersection with the axis is flat, then inhibiting the volatility of the states by increasing λ results in a rather large reduction in dynamic cost and only a slight increase in measurement cost. Conversely, as λ approaches infinity, the states cease to vary and measurement cost becomes large and the slope of the frontier steep. At the margin, the slope of the frontier is thus $-\lambda$.

3.2.2. SOLVING FOR THE LEAST COST TRAJECTORY

From linear control theory, we know that the solution to the cost functional in equation (3.7) will be in the form of a recurrence relation:

$$\phi(b_{t+1}; \lambda, t) = \inf_{b_t} \left\{ r_m^2 + \lambda r_\xi^2 + \phi(b_t; \lambda, t - 1) \right\} \qquad (3.8)$$

where the minimum in the curly brackets is taken over b_t. The initial value of the cost is assumed to be zero — that is, $\phi(b_1; \lambda, 0) = 0$. Again, from results in linear control theory, we know that if the function $\phi(b_t; \lambda, t - 1)$ in the left hand side of (3.8) is quadratic, then the function on the right hand side will also be quadratic. Assume, therefore, that

$$\phi(b_t; \lambda, t - 1) = b_t' Q_{t-1} b_t - 2p_{t-1}' b_t + r_{t-1} \qquad (3.9)$$

---

[1]Schneider (1990) as well as Kalaba & Tesfatsion (1989) have suggested that the cross product matrix $z_t' z_t$ rather than the identity matrix be used as for weighting. I have chosen not to follow their advice.

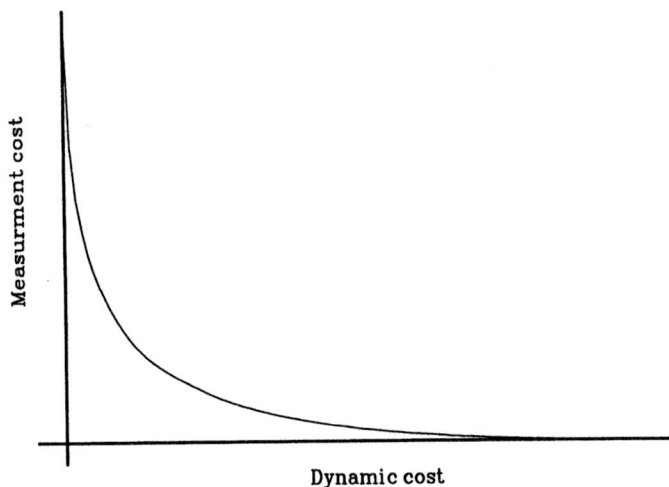

Figure 3.1. **The FLS frontier.** The frontier in the diagram is that for Esab but is typical for all. Note that its slope if flat near the horizontal axis and steep near the vertical. The frontier is convex and reaches its lowest cost when $\lambda = 1$.

Here, Q_{t-1} is a n by n positive semidefinite matrix, p_{t-1} is a n by 1 vector and r_{t-1} is scalar. All three are assumed to have a zero initial value.

Finding the vector b_t that minimizes (3.8) is done by simply collecting the quadratic and linear terms in b_t — using the assumption embodied in (3.9) — and setting the derivative with respect to b_t equal to the null vector. Thus, expanding (3.8) and using (3.9), we find that the quadratic part is[2]

$$b_t' \left(\lambda \Phi' W \Phi + z_t' z_t + Q_{t-1} \right) b_t$$

The linear part is

$$- \left(2\lambda b_{t+1}' W \Phi + 2y_t' z_t + 2p_{t-1}' \right)$$

Finally, the constant part is what is left over:

$$\lambda b_{t+1}' W b_{t+1} + y_t' y_t + r_{t-1}$$

The derivative, set to zero, yields the necessary conditions for the minimum:

$$2 \left(\lambda \Phi' W \Phi + z_t' z_t + Q_{t-1} \right) b_t - 2 \left(\lambda b_{t+1}' W \Phi + y_t' z_t + p_{t-1}' \right) = 0 \qquad (3.10)$$

[2]Note that the square brackets in the formulas that follow do *not* indicate the modulus of the argument inside them.

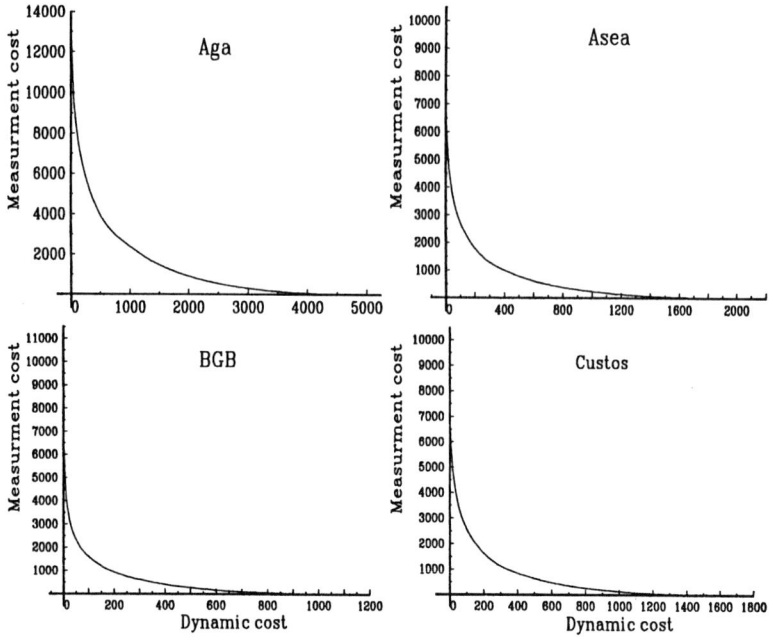

Figure 3.2. **The cost frontier for some of the stocks in the data base.** Note that the scale on the horizontal axis is much greater than that on the vertical. Thus the curves appear flatter than they actually are.

Assuming that the matrix premultiplying the state vector b_t is invertible, equation (3.10) gives us a recurrence relationship from the state at $t + 1$ to the state at t:

$$b_t = \left(\lambda \Phi' W \Phi + z_t' z_t + Q_{t-1} \right)^{-1} \lambda \Phi' W b_{t+1} + k_t \qquad (3.11)$$

where the constant

$$k_t = \left(\lambda \Phi' W \Phi + z_t' z_t + Q_{t-1} \right)^{-1} \left(y_t' z_t + p_{t-1} \right)$$

is also recursively defined.

It is straight forward but tedious to show that Q_t, p_t and r_t may be found by iterating them from their initial value at time 0 through time t.[3]

The formulas for these recursions are also straight forward. We then find that

$$Q_t = \lambda W \left\{ I - \Phi \left(\left[\lambda \Phi' W \Phi + z_t' z_t + Q_{t-1} \right]^{-1} \lambda \Phi' W \right) \right\} \qquad (3.12)$$

---

[3]See, for example, Kalaba and Tesfatsion (1989), p 9-13, for greater detail. The derivation is well known and readily available in the literature. I include the results so that the interested reader can program the routine. A GAUSS program is available on our *ftp* server. See Appendix B.

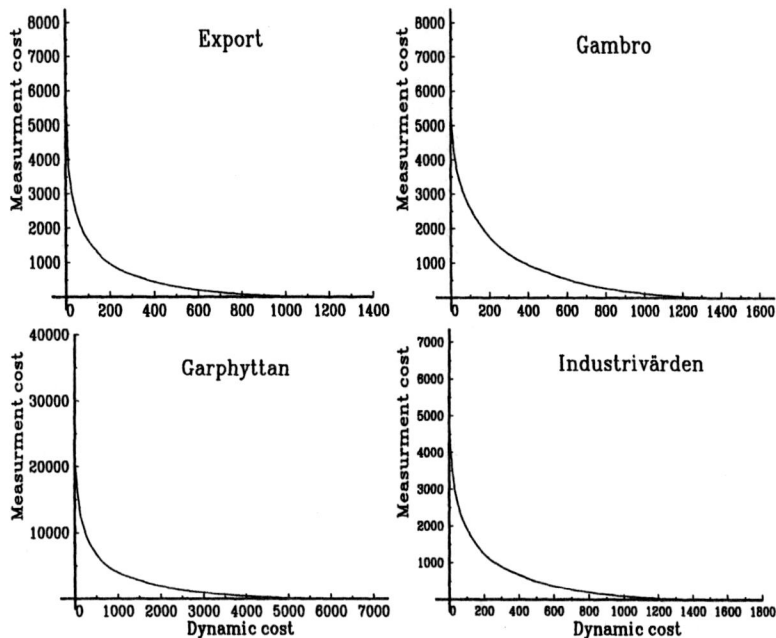

Figure 3.3. The cost frontier for some of the stocks in the data base.

$$p_t = \left(\left[\lambda \Phi' W \Phi + z_t' z_t + Q_{t-1} \right]^{-1} \lambda \Phi' W \right)' \left(z_t' y_t + p_{t-1} \right) \qquad (3.13)$$

$$r_t = r_{t-1} + y_t' y_t - k_t' \left(\lambda \Phi' W \Phi + z_t' z_t + Q_{t-1} \right)^{-1} k_t \qquad (3.14)$$

Given the solution to the above three equations, the final estimate of the state b_T is

$$b_T = \left(z_T' z_T + Q_{T-1} \right)^{-1} \left(z_T' y_T + p_{T-1} \right) \qquad (3.15)$$

Finally, using equation (3.15), equation (3.11) may be used to find the trajectory for state variable. Notice that the state is iterated *backwards* from the final time period. Once this trajectory is found, the minimal cost given λ and W is readily calculated from (3.7). The frontier in Figure 3.1 is also directly available.

What can we conclude for the above exercise? First of all, if the modeling cost is about the same for small values if λ — and by small I mean less than one — but increases greatly as λ grows past one, then a time–varying regression coefficient best describes the data. If, however, the cost remains about constant even as λ becomes as large as 100, then a stable coefficient model is perhaps acceptable. Secondly, examining the time path of the states for differing λ's may also suggest if a stable or a time–varying

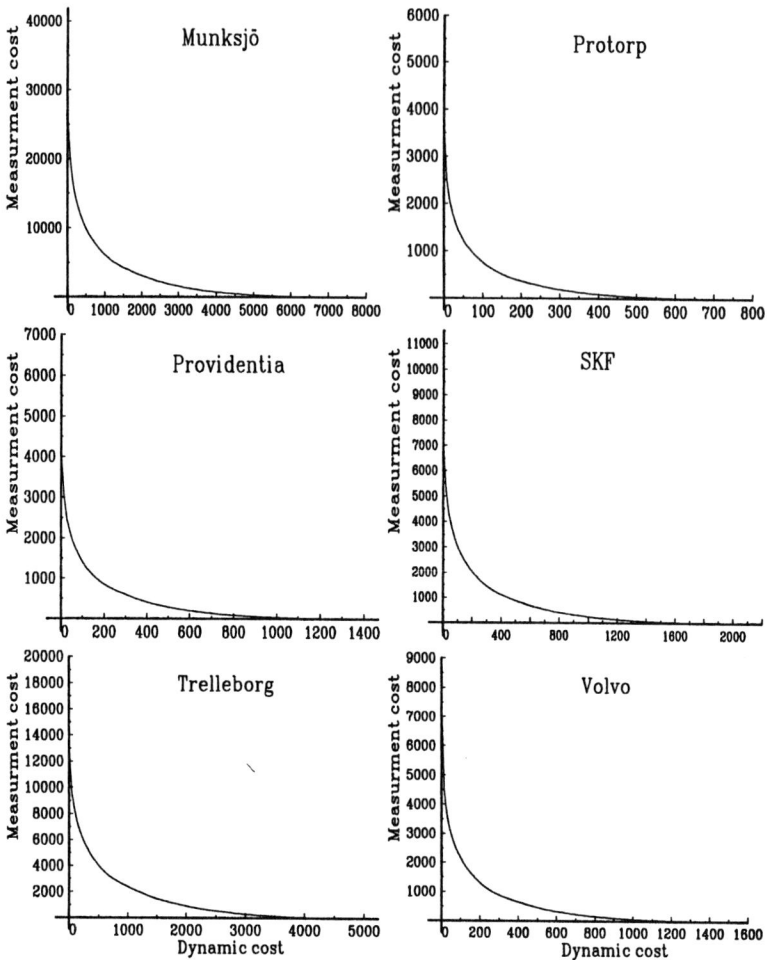

Figure 3.4.   The cost frontier for some of the stocks in the data base.

coefficient model should be used. In particular, if the states vary in spite of large λ's, a fixed coefficient model is more than likely not applicable.[4]

A third interpretation of the method comes from estimating the mean and standard deviation of the states for the different λ values. That these estimates approach the *OLS* estimation as λ increases is obvious. However, the rate of convergence — loosely speaking, of course — indicates the importance of estimating time–varying coefficients. If the average state is close to the *OLS* coefficient value for say $\lambda = 1$, then perhaps the *OLS* model will suffice. The standard deviation of the average state (for given

---

[4]Relative terms like "small" and "large" must be understood in connection with a specific example.

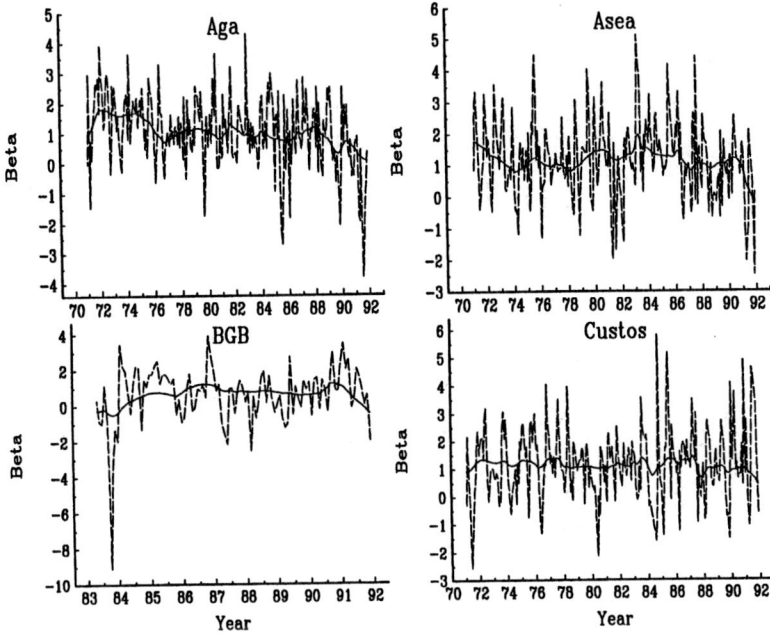

Figure 3.5.   **FLS estimates.** Here we present some *FLS* estimates for $\lambda = 1$ — the broken line — and $\lambda = 1000$ — the solid line. Note that in spite of the large penalty placed on deviations from constancy, the estimates still suggest time–varying coefficients.

λ) is another indicator of the importance of not forcing coefficients to a single stationary value. High volatility strongly suggests that time–varying coefficients should be estimated. Some discussion of a few typical *FLS* estimations follow in the next section.

3.3.   Some examples

It is difficult to give a generalized picture for the *FLS* cost frontier for all of the stocks covered in this study. As a rule, the frontier is very steep and becomes almost vertical as λ becomes large. The evidence here is clear: the square of the residual noise dominates the picture and does suggest that there is much to gain by assuming variable coefficients.

The 10 frontiers depicted in Figures 3.2 - 3.4 support this general observation. The first thing about the diagrams that should be noticed is that the horizontal and vertical scales are different. The second aspect of interest is that most of the curves are very steep. Had the scale been the same on both axes the curves would had almost followed these two lines. The third point of interest is that the frontiers are almost identical in shape even if there exist differences in the proximity to the origin. Thus Providentia — in Figure 3.4 — lies closest to the origin while Munksjö — in

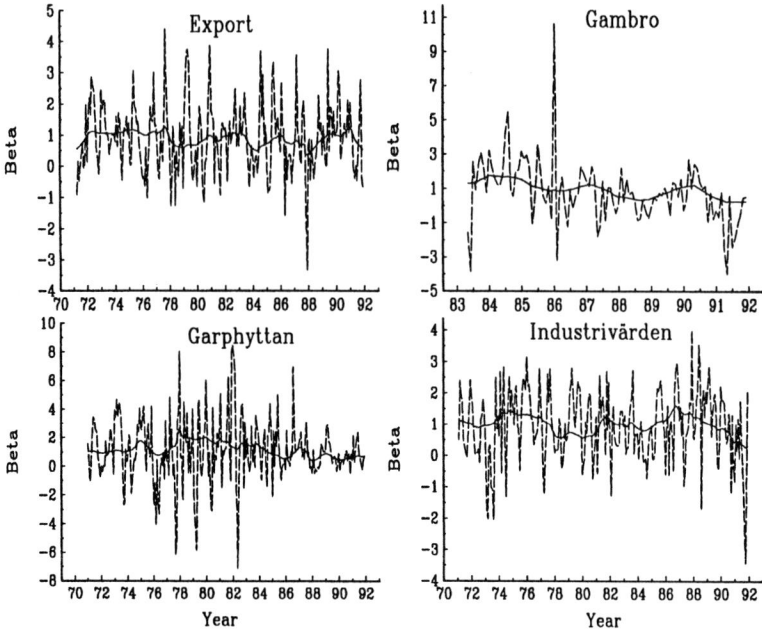

Figure 3.6. More *FLS* estimates. See Figure 3.5.

Figure 3.3 — lies furthest away from it. All the frontiers are flat near the horizontal axis (when $\lambda = 0.0001$) and almost vertical when they intersect the vertical axis (when $\lambda = \infty$).

A further point of interest is that all of the frontiers exhibit minimum cost when λ equals unity. Recall that the slope of the least cost frontier is just $-\lambda$. Thus the minimum cost occurs when the reduction at the margin of measurement cost exactly balances — again at the margin — the increase in dynamic cost.[5]

3.4. A Final Look at the Tests

It is now time to try and summarize the results in the last two chapters. I have made a distinction between constant regression coefficients and constant hyperparameters. A change in the latter, if it occurs at all, is taken to mean a structural shift. Such changes will be discrete rather than continuous: to speak of a structure that is in continuous flux seems to imply a contradiction. A structure as such defines the framework for the model.

[5]As this point is not mentioned by Tesfatsion & Veitch (1990), I am inclined to believe that this minimum is due to the weights applied to the problem analyzed here and not a general characteristic of the method. At any rate, all the frontiers derived by this study exhibit a minimum when $\lambda = 1$.

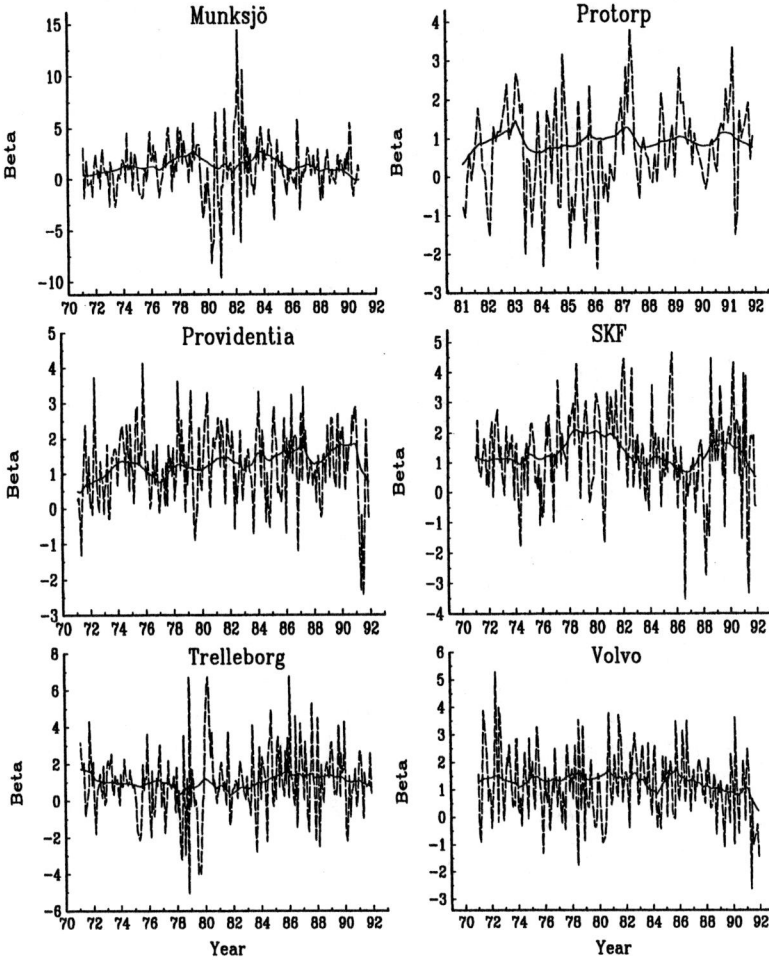

Figure 3.7.   More *FLS* estimates. See Figure 3.5.

This is what the hyperparameters in fact do. They may well change a number of times during the 21 year period studied here, but such changes will occur at definite points in time. Structural change is detected most emphatically by the *cusumsq* test and the *mosumsq* test. Applying these tests to recursive residuals indicates that, at the 10% level, 20 of the 21 stocks show evidence of a structural break using the *cusumsq* test and somewhat fewer — 16 — using the other test.[6] This suggests that the sample should be split into subperiods before estimation. However, there are times when the shift comes at the end of the period: at least 5 stocks enter this category. In this is case there is nothing one can do but wait for future observations to

---

[6]For the entire database of 57 stocks, the corresponding numbers are 50 and 45.

come in. However, as suggested on page 48, it is not always an easy matter to find the exact point where a structural break has occurred. Often there will be two or three points that could be thought of as the first one of the new regime.

While the other tests, in particular the *ARCH* test, are also capable of registering structural breaks, they primarily suggest non constant regression coefficients.[7] As pointed out above, the test results indicate that all of the stocks, showed evidence of variable regression coefficients: the null hypothesis of constancy could not be rejected at the 10% level for at least one of the nine tests presented.

My strategy has been to assume that coefficients are variable if one of the tests were positive. There are two errors that one can make: one could accept the null hypothesis of constancy when reality is in fact variable or reject the null when it was indeed true. I have chosen to consider the first of these two errors as the more serious. Thus I find convincing evidence of non constancy if but one of the tests rejects the null.

My conclusion is, therefore, that, in the data studied here, variable coefficients are the rule. This conclusion is strengthened by the *FLS* plots in Figures 3.5–3.7. With a small weight on the dynamic cost, the *beta* coefficients are very volatile. They vary from 8 units at the smallest — Aga and Asea — to as many as 25 for Munksjö. However, even when the weight is rather large, the coefficient still varies by one or two units. This variation is indeed quite a dramatic variation in systematic risk. Again, the data suggest non constant coefficients.

---

[7]Much of the evidence of *ARCH* disappears when the sample is split into two or more periods. One wonders whether the "evidence" of *ARCH* disturbances in stock returns is really nothing but discrete shifts in process variances. This hypothesis is hard to test as the *cusumsq* statistic is usually not amongst the diagnostic tests presented in these articles.

Chapter 4

The Kalman filter

4.1. A Description of the Kalman Filter

The second topic covered in this book will be the Kalman filter. As an analytical tool, it has been around since the early 1960's when Kalman introduced the method as a different approach to statistical prediction and filtering. The problem addressed is that of estimating the state of a noisy system. For example, consider monitoring the position and the speed of an orbiting vehicle. Rather well known physical laws may be used to describe the system, but at any give moment of time, the exact position and speed of the vehicle may vary from that predicted by the model as there are always external forces at work that result in random impulses to the system. Again, consider the optimal control of an economic system which is described by estimated equations rather than exact physical laws. There will of course be a difference between the theoretical or estimated output from the model and the actual value of the economic variable due to the errors inherent in the model. To be more precise, consider estimating the marginal propensity to save (*mps*) out of disposable income. While it is possible to find a trend or steady state value for the *mps*, its actual value any given year may well deviate from this trend value. In both instances, a well defined concept may not be able to be observed exactly due to the random inputs into the system. Finally, consider the rather simple problem of estimating the time–varying mean of a stochastic system. The mean is assumed to follow a random walk which may or may not contain a deterministic parts. From historical data, an estimate of the mean at the current time is readily available. This estimate may be improved as additional observations of the process become available. Common to all four situations is the that the exact value of the variable studied is not observable and must be estimated.

Engineers need to monitor a system continuously: if their aim is to keep the orbiting vehicle on a given path, they must have estimates of the position and speed of the vehicle and these estimates must be continuously updated as new observations reach them. Such on line monitoring of systems is done by "filtering" the observation to remove the random component so that measures taken to correct the path of the vehicle are based on the "true" or unobserved path. The system studied is as follows:

$$x_t = \Phi x_{t-1} + v_t \tag{4.1}$$

Equation (4.1) is known as the transition equation; the vector x_t is the state vector of the system; its dimension is k by 1. This vector contains information of about the system at time t. Φ is a k by k matrix known as the *transition* matrix. v_t is the random disturbance to the system and is assumed to be normally distributed with a mean of zero and a covariance matrix of Ξ. It is further assumed to be uncorrelated in time. As the equation is a first order stochastic difference equation, it requires starting values. Thus the initial state, x_0, as well as its covariance matrix, P_0, is assumed to be known. We note that (4.1) is not the most general form of the transition equation. It is, however, the formulation used for our estimations. For a more general coverage, the reader is referred to the literature.

The transition equation is the first part of the model; the second is the observation equation which relates the state to the output or observation of the system:

$$y_t = C_t x_t + \varepsilon_t \tag{4.2}$$

In equation (4.2), y_t is the scalar output at time t. C_t is 1 by k and ε_t is a scalar disturbance which is taken to be normally distributed with a zero mean and a covariance of σ^2. Even this error term is assumed to be serially uncorrelated. Further, the independence of v_t and ε_t is also assumed. As suggested by Kalman (1963), the term $C_t x_t$ may be heuristically thought of as the "message" of the system and y_t as the "signal": thus signal consists of the message plus noise. Note that as the state equation (4.1) contains an error term, even the message is stochastic.[1]

---

[1]Here I should point out that the system (4.1)–(4.2) is not the only form that the transition equation may take. For example, many authors — especially control engineers — prefer the following equation:

$$x_{t+1} = \Phi x_t + v_t \tag{4.1a}$$

In this case, the error impacts on the state in the next period rather than the current one. In many physical systems, the period is so short that there is virtually no difference between (4.1) and (4.1a). In economic systems, a good case can be made for preferring (4.1a) to (4.1) if the data are daily or perhaps even weekly: it takes time for the economic agents to react to new signals. If the data are sampled monthly, quarterly or annually, then (4.1) would seem the logical choice. There is, unfortunately, no consensus in the literature.

(4.1) and (4.2) may represent a physical or economic system. However, another interpretation is possible. If we interpret C_t as a matrix of independent variables, then (4.2) could be a regression equation. In this case, y_t is the independent variable and the state vector x_t then becomes a vector of time–varying regression coefficients whose time path is described by the transition equation (4.1). Note that the relationships in the models are not exact: the regression model contains an error term that should be familiar to economists. We also assume that the model for the regression coefficients themselves is not exact. However, whatever our interpretation of the system, we wish to reconstruct the states so that our estimate of the output lies close to the underlying true values of this variables. The Kalman filter is an algorithm that provides such estimates.

There are several derivations of the filter available in the literature. Harvey deduces it from the properties of a multivariate normal distribution (Harvey (1989), pp. 109–110). Kalman himself used the idea of orthogonal projections (see Kalman, 1960) while a number of alternative derivations are given in Jazwinski (1970). Perhaps the least relevant but certainly the most direct of his presentations assumes that the optimal filter is linear and precedes to show that the filter he suggests "with considerable hindsight" (Jazwinski (1970), p. 208) is indeed the one that solves the problem. Thus, assume a linear estimator given by

$$\hat{x}_{t|t} = \hat{x}_{t|t-1} + K_t \left(y_t - C_t \hat{x}_{t|t-1} \right) \tag{4.3}$$

The term $\hat{x}_{t|s}$ denotes the estimate of the state at time t conditional on the information available at time s. The matrix K_t is known as the Kalman

There is, however, not all that great a difference between the two formulations as may at first seem. As state variables are not unique, a system formulated as (4.1) may be transformed to one similar to (4.1a). For example, beginning with (4.1), define a new state vector as

$$q_t = x_t - v_t$$

The system then becomes

$$
\begin{aligned}
q_t &= \Phi q_{t-1} + \Phi v_{t-1} &\tag{4.1b} \\
y_t &= C_t q_t + \xi_t &\tag{4.1c}
\end{aligned}
$$

where

$$\xi_t = \varepsilon_t + C_t v_t \tag{4.1d}$$

and

$$\text{var}(\xi_t) = \sigma^2 + C_t \Xi C_t' \tag{4.1e}$$

That is to say, the variance of the error term in the measurement equation is *not* independent of that in the transition equation. The argument below may be presented in these terms: the interested reader is referred to Jazwinski (1970), pp. 209–211, or to Harvey (1989), pp. 112–113 for details. Here we will assume the original model and the independence of the two error terms.

gain. Assume that it has the following structure. Define the estimation error as at time t conditioned again on information available at time s as $x_{t|s}^e$. Of course,

$$x_{t|s}^e = x_t - \hat{x}_{t|t} \tag{4.4}$$

by definition. Finally, define the error covariance at time t, once again conditioned on the system at time s as

$$E\left\{x_{t|s}^e x_{t|s}^{e\,\prime}\right\} \equiv P_{t|s} \tag{4.5}$$

We seek the minimum of this variance. Here is where "considerable hindsight" becomes useful. Let the Kalman filter have the following form:

$$K_t = P_{t|t-1}C_t'\left(C_t P_{t|t-1}C_t' + \Xi\right)^{-1} + \tilde{K}_t' \tag{4.6}$$

Substituting (4.3) and (4.4) into (4.5), with $s = t$, taking expectations and using (4.6) we get

$$P_{t|t} = P_{t|t-1} - K_t C_t P_{t|t-1} + \tilde{K}_t\left(C_t P_{t|t-1}C_t' + \Xi\right)\tilde{K}_t' \tag{4.7}$$

The term in the parentheses in (4.6) is positive as both R and $P_{t|t-1}$ are variances. The smallest value that the entire last term in (4.7) can assume is zero which occurs when $\tilde{K}_t = 0$. But then (4.6) reduces to the Kalman filter.[2]

For the sake of completeness, and running the risk of repetition, I state below the six equations that make up the recursion relationships of the optimal filter given the output equation (4.2) and the transition equation (4.1). Starting from initial values for the state, x_0, and its covariance matrix, P_0, the recursions first form the estimate of the state and its covariance for the first period. At the next iteration, one uses the result from the first period as the starting point. Therefore, treating period $t - 1$ as the initial period, the estimate of the state and its covariance at time t conditional on information available at $t - 1$ is:

$$\hat{x}_{t|t-1} = \Phi\hat{x}_{t-1|t-1} \tag{4.8}$$

$$P_{t|t-1} = \Phi P_{t-1|t-1}\Phi' + \Xi \tag{4.9}$$

When a new observation of the output — and in the case of time–varying parameters — and the corresponding element of C_t become available, we can obtain the *one step ahead prediction error*, v_t, and its variance, f_t, which is a scalar as there is but one output from out system:

$$v_t = y_t - C_t\hat{x}_{t|t-1} \tag{4.10}$$

$$f_t = C_t P_{t|t-1}C_t' + \sigma^2 \tag{4.11}$$

[2]The reader is referred to Chapter 7 in Jazwinski (1970) for greater detail.

Given the new information, the state and its covariance can now be updated. These are now conditioned on the information available at t.

$$\hat{x}_{t|t} = \hat{x}_{t|t-1} + K_t v_t \qquad (4.12)$$
$$P_{t|t} = (I - K_t C_t) P_{t|t-1} \qquad (4.13)$$

where, as above

$$K_t = P_{t|t-1} C_t' \left(C_t P_{t|t-1} C_t' + \Xi \right)^{-1} \qquad (4.14)$$

is the Kalman gain matrix. It is worth pointing out, that in a *time invariant system* where σ^2 and the system matrices Φ, C and Ξ are known, K_t may be calculated off line as it is not dependent upon observations. While the Kalman gain is itself not constant over time, it usually converges rapidly to a *steady state* value which may be used in the filtering recursions (4.8)-(4.13). However, as the systems treated below will all be time–varying, I will not dwell on this aspect of the problem. Rather, I will list some other more or less interesting aspects of the filter.

The first point concerns the assumption that the error terms follow a Gaussian distribution. The Kalman filter yields an optimal solution to the problem of estimating and updating the state variable of the system. If the error terms are Gaussian, then the filter provides an estimate of the state that is the best available. "Best" should be interpreted as the minimum mean square estimator (*MMSE*) as the state vector itself is stochastic. This estimate is the conditional mean as expressed in equation (4.12). Indeed, the discussion above leads to this conclusion.[3] However, if the error distribution is not Gaussian, then the filter still yields an optimal estimate if we restrict our attention to the class of linear estimators such as (4.12). In this case we speak of a minimum mean square linear estimator (*MMSLE*).

A second point worth noting is that both the *MMSE* and *MMSLE* are unbiased. That is to say, the expected value of the estimate in (4.12) is the true state. The covariance of this estimator is given by $P_{t|t}$ in (4.13). Naturally, both estimates are conditioned by the information available up to and including time t.

To see the next point, substitute (4.10) into (4.12):

$$\hat{x}_{t|t} = (I - K_t C_t) \hat{x}_{t|t-1} + K_t y_t \qquad (4.15)$$

The updated estimate of the state consists of two parts: one that depends on the estimate conditional on information available the previous period and one that depends on the new information available.[4]

---

[3] A more direct proof is given in Chapters 2 and 5 of Anderson & Moore (1979).

[4] Given a time–invariant filter, this dichotomy is clear cut as C_t is then a constant. This is in contrast to a time–varying system where the old estimate is adjusted using both the filter and the new observations on the elements of C_t.

Writing (4.9) at time $t + 1$, and substituting (4.13), yields a nonlinear difference equation known as the *Riccati equation*:

$$P_{t+1|t} = \Phi \left(P_{t|t-1} - K_t C_t P_{t|t-1} \right) \Phi' + \Xi \qquad (4.16)$$

It is interesting to note that the solution to the deterministic linear optimal control problem yields an identical equation. Indeed, it can be shown (see Kalman, 1960) that the problem at hand — finding the *MMSLE* — is the dual to this deterministic control problem.[5]

The above equations are rather abstract. Let us then present a simple example which will illustrate the principles involved.

4.2. An example

Let us return to the example of estimating the time varying mean of a system. Let x_t be the true mean of the system at time t and let y_t be the observation made at that time. Assume now that the observation is rather far removed from the mean value x_t. Two explanations come to mind. First, it may be the case that some unknown force has been applied to the system resulting in the large random deviation which is temporary. The second explanation is that the mean has shifted and the change is of a permanent nature. The filter we apply to the data should ideally not react at all in the first case and shift the system directly to the new value of the mean in the second. If we act on the assumption that the disturbance is random while in fact the mean has shifted, the system will take quite a while to adjust. Conversely, if we over react and assume that a shift has occurred when in fact the disturbance was but a temporary impulse, the system will spend quite a while readjusting to its original mean level. The filter we apply must take both these possibilities into consideration.

The following two equations model this situation:

$$y_t = x_t + \varepsilon_t \qquad (4.17)$$
$$x_t = x_{t-1} + \xi_t \qquad (4.18)$$

In equation (4.17), the signal y_t consists of the message — the mean, x_t — and a random disturbance, ε_t, which is assumed to normally distributed with a zero mean and a variance of σ^2. y_t should be thought of as the physical observation of the mean value of the variable studied. Due to random disturbances, the mean cannot be observed or measured exactly. However, filtering the data will give us an estimate of this unobservable quantity. In (4.18), the mean of the system is assumed to evolve as a

---

[5]On a more personal level, my dissertation Wells (1978) which is mainly concerned with the optimal control problem, also points out this duality.

random walk. The random variate in the equation, ξ_t, is assumed to be normally distributed with a zero mean and a variance q; ξ_t is also assumed to be independent of ε_t. In addition, we assume that the distribution of the initial value of the mean, x_0, is known and normal with a mean of x_0 and a variance of q_0.

The filter is now applied in two phases: first we estimate or predict the next value of the state and the variance of the state:

$$\hat{x}_{t|t-1} = \hat{x}_{t-1|t-1} \tag{4.19}$$
$$P_{t|t-1} = P_{t-1|t-1} + q \tag{4.20}$$

These predictions use the information available before the observation at time t is known. As the state is assumed to follow a random walk according to (4.18), the best estimate of the current observation on the state is its value at time $t - 1$. The variance of this estimate is simply the sum of the variance of the state at $t - 1$, $P_{t-1|t-1}$, and the variance of the random disturbance occurring at time t, q, as these two terms are independent.

The second step in the filtering process is the updating step where the current estimates are revised to take into consideration the new information now available:

$$\hat{x}_{t|t} = \hat{x}_{t|t-1} + k_t \left(y_t - \hat{x}_{t|t-1} \right) \tag{4.21}$$
$$P_{t|t} = (1 - k_t) P_{t|t-1} \tag{4.22}$$
$$k_t = \frac{P_{t|t-1}}{P_{t|t-1} + \sigma^2} \tag{4.23}$$

Equation (4.23) is the Kalman filter. The denominator in the fraction on the left hand side of the equation is the variance of the error in the one step ahead estimate of the mean, $y_t - \hat{x}_{t|t-1}$; the numerator is the variance of the estimated state from equation (4.20). After the state is updated by (4.21), one may improve upon the estimate of the mean by using the new information: that is, a better estimate is $\hat{x}_{t|t}$ in (4.21). As this equation is a rather important step in the filtering process, we should perhaps study it rather carefully.

Rearranging the terms on the right hand side of the equation yields:

$$\hat{x}_{t|t} = k_t y_t + (1 - k_t) \hat{x}_{t|t-1} \tag{4.24}$$

The updated state in (4.24) is expressed as a weighted average of the new observation y_t and the estimate of the state made before the new information became available, $\hat{x}_{t|t-1}$. In time series analysis, the updating is called *exponential smoothing.*[6] Typical, however, for this type of smoothing is that the term k_t is chosen as a constant. If its value is small — in the neighborhood of zero — then relatively larger weight is given to past observations; if k_t is chosen close to unity, then the most recent observation is assumed to be more important. In equation (4.23), this term is not a constant but chosen according to the iterations implied in equation (4.20). Substituting this latter equation into (4.23) we get

$$k_t = \frac{P_{t-1|t-1} + q}{P_{t-1|t-1} + q + \sigma^2} \tag{4.25}$$

k_t is of course less than one; it is also a combination of the variance of the random process that drives the state, q, the variance of the observation error, σ^2, and the estimated variance of the state, $P_{t-1|t-1}$. It is directly proportional to the state variance and inversely proportional the observation error variance. Thus, the more uncertain the state, the greater the weight of the current observation in the revised or updated estimate of the state. Conversely, the more uncertain the observation is, the smaller is the weight applied to the current observation.

Note that the term $P_{t-1|t-1}$ is itself a complicated combination of the other two variances and the initial state variance q_0. Given this initial variance, the state variance may be iterated directly from the updating equation for the state variance (4.22). We may obtain an estimate of the variance of the updated state by substituting (4.20) into (4.22) and (4.23) and, using (4.25):

$$P_{t|t} = \frac{\sigma^2 \left(P_{t-1|t-1} + q \right)}{P_{t-1|t-1} + q + \sigma^2} \tag{4.26}$$

Equation (4.26), given of course an initial value q_0, yields directly an estimate of the state variance for each period.

For time invariant systems such as equations (4.17) - (4.18) above, steady state values of the state variance $P_{t|t}$ as well as the smoothing coefficient k_t — called the *Kalman gain* in the technical literature — have steady state values. To solve for these values, set, in (4.26), $P_{t|t} = P_{t-1|t-1}$ and find

$$\lim_{t \to \infty} P_{t|t} = \frac{q}{2} \left(\sqrt{1 + 4\frac{\sigma^2}{q}} - 1 \right) \tag{4.27}$$

[6]It is worth noting that Friedman, in his study of consumption behavior, suggested exponential smoothing as a technique that could be used to divide consumption data into a permanent and temporary part. An estimate of the permanent part would correspond to the estimated state variable in equation (4.7). The temporary component would then be the residual formed by subtracting this estimated value from the actual observation. See Friedman (1957) and Wells (1992).

Note that an increase in either of the variances will increase the steady state value of the estimated state variance. We can also get a steady state solution to the Riccati equation by substituting (4.27) into (4.20), written at time $t + 1$, and finding

$$\lim_{t \to \infty} P_{t+1|t} = \frac{q}{2} \left(\sqrt{1 + 4\frac{\sigma^2}{q}} + 1 \right) \tag{4.28}$$

The steady state value of the Riccati equation is similar to that of the variance. Along the same lines, we may also calculate the steady state solution of the Kalman gain by substituting (4.27) into (4.25):

$$\lim_{t \to \infty} k_t = 2 \left(\sqrt{1 + 4\frac{\sigma^2}{q}} + 1 \right)^{-1} \tag{4.29}$$

Note that the steady state Kalman gain is inversely proportional to the ratio of the variance of the observation disturbance to the variance of the system disturbance. *Ceteris paribus*, the larger the uncertainty connected with the system itself, the larger is the Kalman gain factor. This in turn implies that when we update the estimation of the state, we should pay more attention to the current observation than to the estimate of the state made using past information. The converse also follows: if the observation is noisier than the system, less weight should be placed on the current observation and more on the estimate made before the new observation became available.

To illustrate the principles discussed above, the system (4.18)–(4.19) was simulated with $\sigma^2 = 0.1$ and $q = 0.05$. The system was shifted upwards in period 100 by 1. Using the y_t thus generated, the "data" was filtered by three different terms: the Kalman gain filter was compared to a "low" feedback as well as to a "high" one. Substituting the variances used into (4.29) gives a steady state Kalman gain of 0.5. After 5 iterations the non-steady state value of k_t, (4.7) is 0.5013; after 8 iterations it is equal to 0.50002; after 10 its value is 0.500001: thus convergence to the steady state value is quite rapid. The steady state solution to the Riccati equation is 0.1 and the steady state variance of the state variable is simply q or 0.05.

For comparison, two other gains were applied: a low-gain of 0.05 and a high-gain of 0.95. Figures 4.1 and 4.2 present the simulations. Note that the low-gain simulation — the left-hand diagram in Figure 4.2 — is always playing a game of "catch up": as the observation falls, the estimate lies above it and vice versa. This is because the estimated system reacts very slowly to new information as the historical value of the system dominates the updating equation. On the other hand, the high gain simulation in

Kalman gain

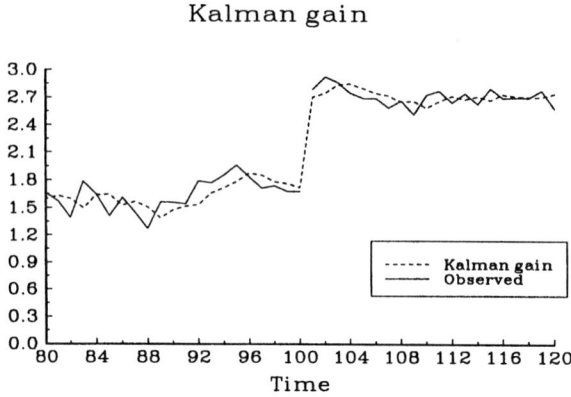

Figure 4.1. **The optimal gain.** The solid line is the simulated data; the dotted line is the estimate obtained from the Kalman filter.

Figure 4.2 over-compensates: the estimate follows the observation rather closely but with a one period lag. Here the system reacts too quickly to new information. The Kalman gain — Figure 4.1 — is lies between these extremes. Indeed, the Kalman gain is optimal in the sense that it provides a minimal mean square linear estimate of the state.

It can be demonstrated that the updated estimate of the state obtained in equation (4.5) minimizes

$$\left(\hat{x}_{t|t} - x_t\right)' D \left(\hat{x}_{t|t} - x_t\right) \tag{4.30}$$

where D is a positive definite weighting factor. In the simulations above, using an identity matrix for D, the sum of the squared deviations of the estimated mean from the actual mean is 2.257 for the Kalman filter; for the low gain filter the corresponding sum is 13.428 and for the high gain 4.645.

An interesting special case arises when the state is observed without error implying that $q = 0$.[7] Then the transition equation becomes $x_t = x_{t-1}$ with the given initial value x_0 and its variance P_0. The measurement equation (4.18) is unchanged. The problem is one of estimating a constant from noisy observations. Even the initial estimate x_0 is uncertain. In this case it becomes possible to obtain an analytical solution to the Riccati equation. Iterate (4.20) one step forward, substitute (4.22) and (4.23) and then invert. At the same time, note that there is no difference between this expression and that for $P_{t|t}$: therefore the expression to the right of the

---

[7]This example is taken from Bryson & Ho (1979).

Low gain filter   High gain filter

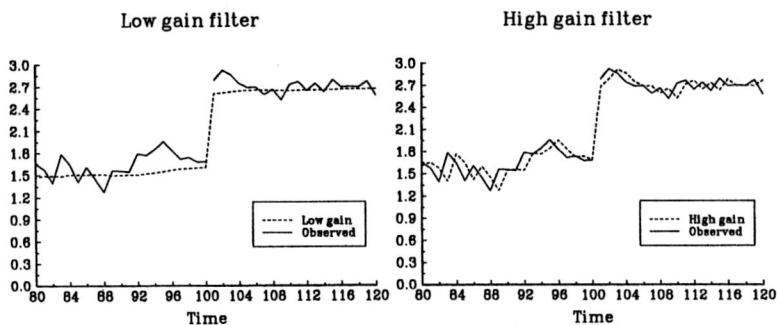

Figure 4.2.   The suboptimal gains. The solid line is the same simulated data as
in Figure 4.1. The dotted lines are the estimates derived using a low-gain filter with
$k_t = 0.05$ (the left-hand diagram) and a high-gain filter with $k_t = 0.95$ (the right-hand
diagram).

horizontal bar is dropped:

$$\frac{1}{P_{t+1}} = \frac{1}{\sigma^2} + \frac{1}{P_t} \tag{4.31}$$

This equation is linear in the inverse of variance of the state. Its solution
is rather straight forward:

$$\frac{1}{P_t} = \frac{t}{\sigma^2} + \frac{1}{P_0} \tag{4.32}$$

or, inverting once more

$$P_t = \frac{P_0 \sigma^2}{P_0 t + \sigma^2} \tag{4.33}$$

As t grows without bound, the state variance shrinks towards zero. The
Kalman gain also goes towards zero as time tends to infinity:

$$k_t = \frac{P_0/\sigma^2}{(P_0/\sigma^2) \cdot t + 1} \tag{4.34}$$

Finally, repeated substitution into (4.21) using (4.34) yields

$$\hat{x}_t = \frac{1}{(P_0/\sigma^2) \cdot t + 1} \left(\hat{x}_0 + \frac{P_0}{\sigma^2} \cdot \sum_{j=1}^{t} y_j \right) \tag{4.35}$$

Note that, from (4.33), the coefficient in front of the sum is simply P_t/σ^2.
From (4.33) we also note that as t grows, the state variance goes to zero.

We may, of course, write zero as $1/t$ for large values of t. Therefore, as $t \to \infty$,

$$\hat{x}_t = \frac{1}{t} \cdot \sum_{j=1}^{t} y_j \qquad (4.36)$$

with a variance of zero. What we have shown is that the mean of the observed values is the best estimator of the unknown constant as the sample becomes very large. The initial estimate, while affecting the estimate for small samples, plays no part in the estimation when the sample is large: x_0 in (4.35) is multiplied by a term that tends to zero as t grows without bound.

A few further items of interest may be noted before progressing. First, assuming that, as above, the variances of the random variables y_t and x_t are known, both the variance of the state estimates as well as the Kalman gain itself may be calculated without reference to the data. This means that the filter coefficients may be calculated off-line and later on be applied as needed. Secondly, the rapid convergence of the gain to its steady state value is partly due to the simplicity of the system; however, even with larger system, the Kalman gain converges as long as the system is time invariant. The project at hand studies *time variant* systems and thus cannot utilize this convergence property. Thirdly, the noise covariances above were considered to be known. While this presents little problems for engineers whose systems follow more or less known dynamic laws and where experimentation may produce reliable estimates, economists have little knowledge of the exact process that drives the system. While one may compensate somewhat for this lack of accurate models by increasing the variance of the system error (see Athens, 1972), this *ad hoc* approach of assigning values to rather important parameters is not appealing: unknown parameters should be estimated with the information available. Indeed one does suspect that other economists share my prejudice and that this could explain why the Kalman filter didn't really become part of the practicing econometrician's tool kit until the mid-1980's.[8]

4.3. Beta and the Kalman filter

The relevance of the above exposition to the problem of the estimation of systematic risk should be apparent. It is therefore not surprising that

---

[8]Of course, the advent of the personal computer may have something to do with it as well. The marginal cost for computing the filter, at least on systems as small as those considered in this book, is essentially zero compared to the all too positive cost in the 1970's when all calculations were done on the University's mainframe. There is, naturally the time cost, but then that was also real in the past when jobs could take days on low priority runs.

a number of authors have used it in their expositions on beta. The first applications seem to have been Kantor (1971) and Fisher (1971). However, while these papers are unavailable locally, an article by Szeto (1973) refers to the latter and even uses the methodology that Fisher suggests. The model proposed for *beta* was the random walk model. Their main problem was estimating the unknown variances in the observation and the system equations. Rather than estimate a constant variance for the observation equation, they used the *OLS* residuals as an estimate of the error sequence. Then the variance of the observation equation was calculated using different time lengths. That is, the variance at period t was based on the residuals from the first to the t'th period. The variance of the system noise was model more problematical. They took this variance to be proportional to that of the observation equation. Thus the factor of proportionality was the ratio between the *OLS* estimate of the variance of the estimated *beta* and the *OLS* residual variance for the entire sample. Thus, assuming that this ratio was constant, they could also generate a sequence of variance terms for the system equation. Rosenberg (1973) mentions *beta* and also presents a maximum likelihood method for estimating the unknown variances. His main interest in this paper was, however, estimation and *beta* appeared only in passing.

Garbade & Rentzler (1981) use the Kalman filter to estimate the variance of the system equation (4.1) when a random walk model is assumed. They present the likelihood function to be maximized, and the likelihood ratio test that could be used to test the hypothesis of *beta* stationarity. Ohlson & Rosenberg (1982) also use the filter and likelihood ratio tests in their study of the equally weighted CRSP stock average for securities on the New York Stock exchange. Their comment on the various models is relevant: "The most crucial improvement in goodness of fit is obtained by allowing for variation in *beta*, whether random or sequential". (Ohlson & Rosenberg (1982), p. 132)

Bos & Newbold (1984) used the Kalman filter to estimate systematic risk as modeled in (4.1) with the coefficient Φ restricted to the interval of minus one to plus one. Their state vector was univariate so the transition matrix is but a single coefficient, Φ_{11}. They conclude that the evidence indicates a non constant *beta* even if the exact functional form cannot be pinpointed. On page 21 I commented on the sign of Φ_{11} but the discussion is worth repeating. If this coefficient is negative and less than unity, the time path of the systematic risk alternates between positive and negative values, even if the fluctuations are dampened. With Φ_{11} between zero and plus one, *beta* adjusts smoothly towards its constant value. In particular, control engineers reject the former model; indeed, a negative Φ_{11} may be confused with the random coefficients model. Collins *et al.* (1987) also discuss this point and conclude that Φ_{11} should be restricted to the interval between zero and one.

Fisher & Kamin (1985) use the filter without using maximum likelihood estimations. They assume that var(ε_t) is heteroscedastic and proceed to derive estimates for *beta*. They also use a slightly different output equation which obtains by removing the means of the two variables and then by dividing through the equation with the independent variable — that is, with the market index. However, empirical results are not presented: they refer the reader to an unpublished study which "improves on the estimation of *beta*". (Fisher & Kamin (1985),p. 140)

Finally, I note the work of Knif (1988, 1989) who estimated systematic risk on the Helsinki stock market using the Kalman filter. While he refrains from conclusion as to the pattern of this time-variation in *beta*, his estimates show his efforts to be justified. A preliminary abstract of this book has also been published (see Wells, 1994). The main differences between my findings in that paper and those presented later on are mainly due to the extension of the data base.

Chapter 5

Parameter estimation

5.1. Preamble

The Kalman filter is has long been a standard tool for control engineers. However, the initial introduction of the filter to an audience of economists, while emphasizing the relevance of the filter in modeling economic systems, also pointed out the need to *assume* known covariance matrices for the various noise processes in the model. In an article in the *Annals of Economic and Social Measurement*, Athens (1972), after pointing out that the noise parameter in the system equation could represent "input uncertainties and deterministic modeling errors", continues by saying:

> "Thus the covariance [of the error term in the system equation that is] *selected by the designer* should incorporate his judgment on the importance of the higher order terms in the validity of the linearized model. Thus, the 'more nonlinear' the system dynamics, the 'larger' the [covariance] should be. The white noise ... in the observation equation plays a similar role. Not only should it reflect the inherent uncertainty of the measurements due to sensor inaccuracies, but it should also be used to model the implications of neglecting [higher order terms] to obtain a linear equation." (Athens (1972), p. 472, emphasis added)

Error terms are meant to capture the effects of a poorly specified model; as Athens points out, they allow us to cheat by studying linear models. He continues:

> "This is extremely important because as we shall see ... we shall ask some very precise questions of the mathematics. If we ask precise but stupid questions, we will get precise but stupid answers."

Here is the essence of the problem: social scientists do not know the degree of stupidity embodied in the question. That is to say, they do not know how poor — or how good — the specification actually is. One can

89

always launch another satellite in the same path as the former in case something really goes wrong just to gather another set of measurements. This course of action is not open to economists: one cannot rewind the economy and have another go at it in order to get a better idea as to the size of the errors involved.[1] One may naturally balance uncertain observations against an approximate model by the astute choice of covariances. Referring back to the example in Chapter 4, the steady–state filter derived there in equation (4.29) exemplifies this choice. If the measurement uncertainties are larger than the "modeling" ones, then less weight is placed on the current observation. However, if we are rather sure that what we actually observe is indeed the true value, then this term is weighted more heavily than the uncertain estimate of the state. We have also seen how different weights may be used to model uncertainty in our discussion of the flexible least squares technique. But the fact remains that important parameters are being assigned *a priori* values rather than ones somehow more closely related to the data. Indeed, the trade–off between the measurement and the system noise requires some index. The flexible least squares used a "cost" index where measurement cost was measured as the sum of the squared estimated deviations from the actual measurements — equation (3.3) — and the dynamic or system cost was the volatility of the state — equation (3.4). We saw that this cost was smallest when equal weights were placed on the cost terms. Still, this is no great help to us if we know nothing about the relative magnitudes of the covariances.

Rosenberg (1973) — in the same journal that had published Athens' paper — presented a maximum likelihood method for estimating the unknown parameters in the filter. While Rosenberg was mostly concerned with finding estimates that "converged" to the "true" and constant population norm, he outlined the methodology that we use here. We did not, however, see a flux of journal articles estimating time–varying parameters: the norm was constant parameters and estimates of them were sought through more or less complicated statistical procedures. The cost of computer time was surely a factor in this development. It is also interesting to note that Athens, in the *Annals* for 1974, presents the Kalman filter once again. However, he merely assumes that the covariance of the error terms had "been estimated" or were somehow "known".

---

[1] Others, of course, face closely related problems. It has been suggested, for example, that by pumping water into the San Andreas fault — that's what threatens to dump California into the Pacific Ocean — one could entice the opposing plates to glide smoothly past each other, rather than building up pressures and then moving all at once. The problem is, of course, no one is really sure what would happen. Perhaps the water would produce an extremely large earthquake and, once that happened, you could not fish California back out of the sea. It's one thing if God causes earthquakes; it's quite another if the United States Geological Survey causes them.

Garbade (1977) seems to have presented the first widely published application of Kalman filtering to the problem of estimating an econometric model other than the *CAPM* with varying coefficients. Using simulation studies he compares the relative strength of the *cusum* and the *cusumsq* tests against a likelihood ratio test based on the maximization of the likelihood associated with the varying coefficient model. (see below) He also estimates varying coefficients for a monetary model presented by Kahn and is careful to point out the merits of this estimation, especially in illustrating fluctuations in the elasticity of the demand for money with respect to both commercial papers and high powered money. He does note, however, that if the variation in parameters is small then the extra cost of the maximum likelihood estimates is perhaps not justified. He does not, however, explain how one *a priori* can decide on the intensity of the parameter variations. One strategy would be to pretest the data by applying the tests present in Chapter 2. Here the *cusumsq* plots would be especially helpful. The *fls* estimates would add further evidence of parameter stability. If the results indicate non-constancy, then the more expensive — remember that computer time was expensive in the mid 1970's — method could be worth while.

Even if Garbade seems to have been first out of the gate, I associate the application of the filter to parameter estimation with the name of Harvey. Pagan (1980) refers to a discussion paper by Harvey and Phillips dating from 1978.[2] In this paper, the authors present both the state variable representation and the likelihood function for the *MRV* model as well as estimates from a Monte Carlo simulation with a very small number of simulated observations (20). The experiment was repeated about 350 times for different coefficient values. They conclude that there is a "clear gain" from using the maximum likelihood methods compared to *OLS* and a two-step method involving first the estimation of the model parameters — the coefficients in the diagonal transition matrix and the variances of the noise parameters — and then the mean values of the state variables (Harvey & Phillips (1982), p. 318).

Harvey has concentrated his attention to structural models, that is, models that reduce a time series to its basic structure consisting of a trend, cyclical, seasonal and random components. The basic difference between such models and those considered here is that the former are *time invariant* while the latter are not. His results are summarized nicely in a volume Harvey (1989). The basic problem we face is that the variance of error terms in the output and the transition equations as well as the elements of the transition matrix are unknown and must be estimated from the data.

---

[2]The paper was later published. (see Harvey & Phillips, 1982). My discussion is of course based on the published paper.

5.2. Choosing a model

There have been rather many models for systematic risk described in the literature. All of them can be represented by a simple two equation model. The first of these two equations is the *observation equation*:

$$y_t = C_t (x_t + \Gamma) + \varepsilon_t \tag{5.1}$$

Basically, this is the market model with time–varying coefficients. In the general case, the C_t matrix consists of a column of ones and the returns to the market index and is assumed to be known. Its dimension is T by 2, so that each row will represent the observations at a certain point in time. The *state vector*, x_t, is conformal to C_t. With a dimension of 2 by 1, the vector will represent the *alpha* and *beta* coefficients at time t. Γ is a matrix of constant terms that are added to the state at each t. In the work below, Γ will not be time–varying. Its dimension is also 2 by 1; it contains the steady state values of the time–varying model parameters. These may well be zero, so this matrix is not included in all model specifications. ε_t is the part of the return y_t to the asset which is not modeled. The variance of ε_t, σ^2, is unknown and assumed finite.

The returns to the assets are defined as discrete monthly returns:

$$y_t = \left(\frac{P_t + d_t}{P_{t-1}} \right) \cdot 100$$

where P_t is the price of the asset at period t[3] and d_t is any dividend paid out during the month. The market index used is *Affärsvärlden's General-index* of the stocks on the Stockholm exchange, which is a value weighted index of the securities on the Stockholm exchange and is calculated daily.[4] The returns to the market portfolio are also calculated discretely.

The second equation used in describing systematic risk in the market model is one that traces out the path of the time–varying coefficients. This equation is really a first order stochastic difference equation with constant coefficients. In its most general form it appears as follows:

$$x_t = \Phi x_{t-1} + \xi_t \tag{5.2}$$

The matrix Φ is assumed to be diagonal; even the covariance matrix of ξ_t, Ξ, is assumed diagonal and of course finite. The elements in these matrices are unknown. x_t, as explained above, is the vector of the regression

---

[3]The price used is the closing price on the last day of trading during the month in question.

[4]The actual data used is taken from the table published by Hansson & Frennberg (1992) which contains a few "typographical" errors which have been corrected here. The errors are not entirely theirs as the published *Generalindex* contains the same errors. Somewhere along the line, a typo became fact.

coefficients of the system at time t. However, in certain models, this vector will represent the deviation of these coefficients from their respective steady–state values at time t.

Before discussing more explicit model formulations, I return once again to the discussion the elements of Φ, Φ_{ii}, that was begun on pages 21 and 87–88. Repeated substitution yields the state as a function of its initial value and the error components:

$$x_t = \Phi^t x_0 + \sum_{i=1}^{t} \Phi^{t-i} \xi_t \qquad (5.3)$$

x_t should be bounded — a rather reasonable requirement as systematic risk has not been observed to increase without bound. Indeed, there are those who have maintained that this risk should tend to unity, and, even if we have questioned their methods, their findings do indicate that the risk is not unbounded. If, then x_t is bounded, then absolute value of Φ_{ii} must not be greater than unity.

If $\Phi_{ii} = 1$, then each state follows a random walk independent of the other state. The volatility of the state will depend on the its initial variance and, with increasing importance, the sum of the noise covariances through the periods up to and including the present:

$$\text{var}(x_t) = \text{var}(x_0) + t \cdot \Xi \qquad (5.4)$$

Random walks are in one sense bounded: the expectation of the difference in the state between two periods is zero. The time path of the state depends on the variance of the noise that is driving the system: the larger the variance, the more the state will move about.

If $\Phi_{ii} = 0$, then the state is simply a vector of random fluctuations. For this system to have economic meaning, Γ in equation (5.1) will have to be non-zero. The elements of this matrix will then be interpreted as the steady state value of the parameters and the states will be but random deviations from these values.

If Φ_{ii} is non-zero, but not exactly unity, then Γ will also be non-zero and the states once again interpreted as deviations from the steady state values found in Γ. From equation (5.3) we note that, as time progresses, the initial deviation has little influence on the current state. The same may be said about past disturbances. Of course, if Φ_{ii} lies close to one, then past influences die out slowly and the model resembles a random walk about a given mean. In Chapter 4 we considered whether Φ_{ii} could be negative but greater than minus one. The issue may seem trivial — after all, the model parameters will be estimated and if the best estimate yields a negative Φ_{ii} then that's how it is. However, as suggested there, things are not all that simple. If Φ_{ii} is allowed to lie between minus one and zero, then, as equation (5.3) also aptly demonstrates, the deviation of the state from the value given by Γ is positive for even periods and odd for uneven ones. Especially if Φ_{ii} is close to zero but still negative, the time path of the state will be almost indistinguishable from that of $\Phi_{ii} = 0$. Given Φ_{ii} close minus one, fluctuations become even more violent as past disturbances fade slowly. Even if more eminent persons than myself have allowed Φ_{ii} to vary between plus and minus one — notably Bos & Newbold (1984) and Knif (1989) — I have chosen to restrict these parameters to be positive and not greater than unity: $0 \leq \Phi_{ii} \leq 1$.

5.3. Variations on a Theme

Equations (5.1) and (5.2) are the model. All of the models of systematic risk that I have encountered in the literature may be expressed by restricting some of the parameters of the model and by changing the dimension of the state vector. All of the models are nested and indeed collapse to the simple *OLS* estimation of the market model. Most of the variations considered below have already been presented: they are listed here for the reader's convenience.

I will make one general observation before going into details concerning the treatment of the constant in the market equation, α. Many studies subtract the mean for the variables and set $\alpha = 0$ *a priori*. Although the risk free rate of interest or the means could be subtracted before estimation, I have tried as far as possible to estimate a time path for α; however, many of the estimates — as will be seen later on — are rather poor. Indeed, in the *OLS* specification, α is almost always less than twice its standard error so one really does not expect much better results with more complicated methods. As I present the various models below, I will mention how the problem of the constant is approached.

5.3.1. THE RANDOM WALK MODEL

My favorite model is the *random walk* model (*RW*). There is voluminous literature as to whether asset prices follow a random walk. There is even a sizable body of literature beginning with Fisher (1971) and Kantor (1971) that asserts that even *beta* follows a random walk. Sunder (1980) and LaMotte & McWhorter (1978) have developed parametric statistical tests designed to test if *beta* follows this process. Given that the market reacts to firm dependent developments and not just to noise, one may use the path traced by *beta* as a random walk to examine events within the firm that have been reflected in the market's evaluation of the risks involved in holding this asset.[5] To specify this model, one sets $\Gamma = 0$ in equation (5.1) and Φ_{ii} equal to the identity matrix in equation (5.2). The state variables will contain the time paths for both *alpha* and *beta*.

There are three parameters that must be estimated: σ^2 and the diagonal elements of Ξ which are the variances of the stochastic terms. These variances cannot be negative; therefore, the estimates are constrained to be positive. Rather than using constrained optimization techniques, these parameters have been estimated without restriction. When the model, and the likelihood function based upon it, is calculated, the estimates are squared: this guarantees positive variances.

5.3.2. THE RANDOM COEFFICIENTS MODEL

As we have seen, *random coefficients* model, *RCF*, has also been studied for other stock markets. The first mention of this model seems to have been Schaefer *et al.* (1975) although their paper presented in Budapest in 1972 may have mentioned the model.[6] It is also an attractive model implying that there is a "true" or "long run" risk associated with a given company's stock even if temporary disturbances may tend to increase or decrease this risk. However, one would question how the assets of a company that continuously expands into new markets and leaves others could conceivably have a constant long run risk. Discrete shifts could increase the relevance of the model, but the real world seems to be continuous. In any case, this model has been considered here. To derive it from (5.1) and (5.2), we reduce the dimension of the state vector by one as a randomly varying *alpha* cannot be distinguished from the error term in the observation equation. Thus the state variable will be *beta*'s deviation from its mean value. *Alpha* will also be added to the right and side of (5.1). The parameters to be estimated are the variances of ε_t and ξ_t and the mean of the single state which is contained in Γ as well as the constant α.

---

[5]I note here that I am embarking on a dangerous journey. The state representation of *beta* is, by in large, arbitrary. By changing the model one changes the time path of the coefficient — i.e., the history of the risk development of the firm. (See Section 6.3.) This is most unsatisfactory if one believes in the information contained in the state variables and wishes to use them to test hypothesis of the market's reaction to changes — announced or inferred — in company policy. However, I intend to use the estimated states to describe general tendencies and leave to others the problem of developing methods of testing the market's reaction to new information. After all, with monthly data, all news is old news: the market will capitalize on changes with minutes. Thus I cannot ask exact questions or expect exact answers.

[6]Terminology can at times be anything but clarifying. In financial literature, the model that I call the *random coefficients* model presented here is often called a "mean-reverting" model. Schaefer *et al.* call it the "dispersed coefficient" model. As my background is more econometrics than finance, I prefer to follow Pagan's label and call a model whose parameters vary randomly about a mean value a *random coefficients* model.

5.3.3. THE MEAN REVERTING MODEL

The *mean reverting* model, *MRV*, is another model that has been of great interest to many of those who have studied time–varying systematic risk. Rosenberg (1973) seems to have been the first to mention it in the present context. As the number of parameters estimated is large, some have reduced the dimension of the problem by subtracting the mean and the risk free rate of interest from the variables. This approach was suggested by Ohlson & Rosenberg (1982). That *alpha* is often insignificant in the *OLS* estimates is another rationalization for reducing the dimensionality of the problem. On the other hand, Sweden lacked a market determined risk free rate of interest until March 1981. As subtracting a non market determined rate seemed pointless, when the dimensionality of the model has been reduced, this reduction has been accomplished by estimating the intercept as a constant instead of subtracting means and risk free rates before estimation.

If the full model is estimated, there are seven unknown parameters: the three variances, σ^2 and the diagonal elements of Ξ; the two unknown means in Γ; and the diagonal elements of the system matrix Φ. These latter two have been restricted to lie between zero and one by estimating the unconstrained coefficient and the calculating the model parameters using

$$\Phi_{ii} = \frac{\Omega_{ii}^2}{1 + \Omega_{ii}^2} \tag{5.5}$$

where Ω_{ii} is the unrestricted value of the parameter.[7] There are of course other alternative methods of restricting the coefficient, such as using the absolute value of the unrestricted parameter instead of its square, but I have found that (5.5) works rather well. It has also been used by others, notably Rosenberg (1973) and Ohlson & Rosenberg (1982). If the absolute value of the estimated Ω_{ii} is large, the model approaches the random walk model; small absolute Ω_{ii} approximate the random coefficients model. Thus the results of an estimation of this model can suggest another specification. When the restricted model is estimated, the variance of the α factor may of course be ignored thereby reducing the number of parameters needed for estimation from seven to six. "All" that need be estimated in this case is σ^2, one element in Ξ, the two elements in Γ and of course Φ_{22}.

5.3.4. THE OHLSON-ROSENBERG MODEL

This model, the *ROS* model, is the variant used by Ohlson & Rosenberg (1982). It has been described by equations (1.22)-(1.25) in Chapter 1. In

---

[7]The discussion on pages 21, 87–88 and 93–94 presents the rationale behind this restriction.

this model, the state variables are defined as deviations from their steady state value and the variance of the first state is set to zero so that *alpha* is a constant, and need to be added to equation (5.1):

$$y_t = \alpha + C_t (x_t + \Gamma) + \varepsilon_t \qquad (5.6)$$

As there is but one state variable — *beta* — the C_t matrix is one by T. The state equation (5.2) stands as is, but the variance of ε_t is now heteroscedastic: $\text{var}(\varepsilon_t) = C_t^2 \lambda + \sigma^2$. This gives us six parameters that need to be estimated: σ^2, λ, α, and Γ in the observation equation (5.6) as well as Φ_{22} and Ξ_{22} in the state equation. In this model, *beta* behaves as a random coefficient — thus the extra term in the variance of ε_t. But the difference between the actual state — the actual *beta* — and its mean follows a first order autoregressive scheme. Thus, if $\Xi_{22} = 0$ and $\Phi_{22} = 1$, the model becomes the random coefficients model; if on the other hand, $\lambda = 0$, the model becomes the mean reverting one. With $\Phi_{22} = 1$, the model becomes the heteroscedastic one studied by Fisher & Kamin (1985). Finally, with $\Phi_{22} = 1$ and $\lambda = 0$, the model collapses to the random walk model.

5.3.5. THE MOVING MEAN MODEL

Another variant, which is absent from the literature except in a passing comment in my article (see Wells, 1994) in response to a referee's request, is the *moving mean model*, (*MM*). It was presented by equations (1.27) – (1.28) in Chapter 1. In it, *beta* is mean reverting but it reverts to a mean that itself develops as a random walk. This model requires that the state vector have a dimension of three: one state for *alpha*, one for *beta* and one for the moving mean of *beta*. The state equation will then be

$$x_t = \begin{bmatrix} \Phi_{11} & 0 & 0 \\ 0 & \Phi_{22} & 1 - \Phi_{22} \\ 0 & 0 & 1 \end{bmatrix} x_{t-1} + \begin{bmatrix} 1 & 0 & 0 \\ 0 & 1 & 1 \\ 0 & 0 & 1 \end{bmatrix} \begin{bmatrix} \xi_{1t} \\ \xi_{2t} \\ \xi_{3t} \end{bmatrix} \qquad (5.7)$$

The observation equation is in this case:

$$y_t = C_t x_t + \varepsilon_t \qquad (5.8)$$

Now C_t is T by three: a vector of ones followed by the returns to the market index and followed a column of zeros. In one of the estimates presented below, Φ_{11} has been estimated, which requires that six parameters be estimated: the four variances — σ^2 and the diagonals of Ξ — and the two autoregressive parameters — the diagonal of the transition matrix Φ. In most cases, however, *alpha* is estimated as a constant; this reduces the dimension of the state vector as well as the number of estimated parameters by one.

5.4. The likelihood function

We are faced with the problem of estimating the unknown the variances of the error terms and the elements of the transition matrix Φ in the above models. These parameters are called the *hyperparameters*. The data on which the estimations will be based consists of observations of the returns to individual securities and to the market portfolio, *Affärsvärldens Generalindex*. The state vector x_t is in this case a two dimensional column vector $(\alpha_t, \beta_t)'$ while the output y_t is scalar and represents returns to an asset. The C_t matrix in (5.1) has the dimension T by 2 where T is the number of observations used in the estimation.[8] The first column of C_t is a column of ones and the second a column of the returns to the market index. By adding and subtracting $C_t \hat{x}_{t|t-1}$, where, as before, $\hat{x}_{t|t-1}$ indicates the unbiased estimate of the state vector at time t conditional on information available at time $t - 1$, equation (5.1) may be rewritten as

$$y_t = C_t \hat{x}_{t|t-1} + C_t \left(x_t - \hat{x}_{t|t-1} \right) + \varepsilon_t \tag{5.9}$$

Taking expectations conditional upon the information set at $t - 1$, the last two terms vanish as both ε_t and the expression in parentheses have an expected value of zero: the conditional expectation of y_t is thus $C_t \hat{x}_{t|t-1}$.

To derive the variance of y_t, rewrite equation (5.9) as:

$$y_t - C_t \hat{x}_{t|t-1} = C_t \left(x_t - \hat{x}_{t|t-1} \right) + \varepsilon_t$$

The variance of (5.9) is thus easily seen to be :

$$\text{var}(y_t) = \underset{t-1}{E} \left\{ y_t - \hat{y}_t \right\}^2 = C_t P_{t|t-1} C_t' + \sigma^2 \tag{5.10}$$

Here $\hat{y}_t$ indicates its conditional expectation in (5.9); the matrix $P_{t|t-1}$ is the variance of the estimate of the state at time t conditional on the information available at time $t - 1$. The final term in the above equation is of course the variance of the error term ε_t.

At this point we make an assumption on the distributions of ε_t, ξ_t and the initial value of the state variable, x_0. All are assumed to follow a normal — or multivariate normal as the case might be — distribution. The first two of these have a zero expected value; ε_t has a variance of σ^2 and ξ_t a covariance of Ξ. P_0 is the variance of the initial state variable. These assumptions allow us to write the likelihood function in terms of the above variances. In writing this function, we utilize the fact that y_t is scalar and thus the variance in (5.10) is also scalar. v_t is defined in (4.10): it is the

---

[8]The discussion that follows assumes that Γ in (5.1) is zero.

one step ahead prediction error and f_t is its variance; T is the number of observations and k is the number of estimated parameters.

$$L = -0.5\,(T-k)\left(\ln 2\pi + \ln \sigma^2\right) - 0.5 \sum_{t=k+1}^{T} f_t - 0.5\sigma^{-2} \sum_{t=k+1}^{T} \frac{v_t^2}{f_t} \quad (5.11)$$

The parameters that are used to maximize L are the hyperparameters of the model. Note that these may also be a "mean value" in the *RCF*, *MRV* and *MM* models.

As numerical optimization is time consuming, it is advantageous to reduce the number of parameters in the problem. This may be accomplished by the trick of concentrating one of the parameters out of the likelihood function above. The one chosen will be one of the variances as we may partition these parameters into two separate parts, one of which will be scalar. We then express the remaining variances as a function of this one. The obvious choice is the variance of ε_t, σ^2, allowing us to write

$$\mathrm{cov}(\xi_t) = \Xi \equiv \sigma^2 Q$$

As σ^2 does not appear in the formulas for the residuals, v_t, or their variance, f_t, the elimination of σ^2 is accomplished by simply setting the derivative of L with respect to σ^2 equal to zero. This derivative is

$$\frac{\partial L}{\partial \sigma^2} = -\frac{0.5\,(T-k)}{\sigma^2} + \frac{1}{(2\sigma^2)^2} \sum_{t=k+1}^{T} \frac{v_t^2}{f_t} \quad (5.12)$$

Solving (5.12) for σ^2 yields an estimator of the variance of ε_t, $\hat{\sigma}^2$, that is conditional on the other parameters in the likelihood function:

$$\hat{\sigma}^2 = \frac{1}{T-k} \sum_{t=k+1}^{T} \frac{v_t^2}{f_t} \quad (5.13)$$

To obtain the likelihood function that is actually maximized substitute (5.13) into (5.12):

$$L^{\dagger} = -0.5\,(\ln 2\pi + 1) - 0.5\,(T-k)\ln \hat{\sigma}^2 - 0.5 \sum_{t=k+1}^{T} f_t \quad (5.14)$$

Numerical optimization methods have been used on the above function of obtain estimates of the unknown parameters and variances of these estimates. A consistent estimator of the latter has been suggested by White (1982) to be the diagonal elements of

$$\hat{S} = T^{-1}\hat{H}^{-1}\hat{G}\hat{H}^{-1} \quad (5.15)$$

Here $\hat{G}$ is the cross product matrix of the first derivatives of the likelihood function at the final iteration while $\hat{H}$ is the Hessian or the matrix of the second derivatives again at the final iteration. The quality of the estimates should be judged both on the goodness of the fit statistics as well as the diagnostic tests presented below.

5.5. And the Winner is ...

Given the various models and the estimates of them, we must be able to choose which one is "best". This is no easy matter and often it is prejudice that decides the winner. The models are all nested — that is, by placing appropriate restrictions on the most general of the models we obtain the others. For example, by setting $\Phi_{11} = \Phi_{22} = 1$ in the mean reverting model, the random walk models obtains; by setting $\Phi_{11} = \Phi_{22} = 0$, the random coefficients model results. Thus one is tempted to use the coefficient of determination as a measuring stick: the higher R^2 or perhaps the adjusted R^2, the better the fit.

There are, however one or two minor problems with this approach. First of all, we can calculate two R^2's: one for the one-step ahead prediction residuals and one for the residuals formed from the updated states. Which one should be used? If the OLS R^2 is included for comparison, then the coefficient of determination calculated from the one-step ahead residuals is at a disadvantage as one of the two statistics uses more information than the other. However, using the R^2 from the updated states obscures some of the advantages of using the Kalman filter. After all, if one really needs an *ex ante* prediction, then using an *ex post* indicator of model adequacy is not very satisfactory.

Another minor problem concerns the variables used in the estimations. The RCF and the MRV models may be estimated from variables that have had their respective mean — as well as the risk free rate of interest — subtracted before estimation. In these cases, as the variables are not identical, the models are not truly nested. If the models are non-nested, then comparing R^2's is like comparing beer and oranges. The usual way out of this dilemma is to use the *Akaike Information Criterion* (AIC) which is designed to weight the reduction in the likelihood function against the increase in the number of parameters necessary to achieve this reduction. For the models under consideration here, the number of observations is rather large and the AIC may be expressed as follows (Judge *et al.* (1985), p. 245; Harvey (1989), p. 270):

$$AIC = \sigma^2 \exp\left(2\frac{k+s}{T}\right) \tag{5.16}$$

Here σ^2 is the estimated residual variance of the model, s is the dimension of the state vector and k the number of hyperparameters that are to be estimated. In theory, the smaller the *AIC*, the better the model. However, this criterion is rather *ad hoc*. There is no standard error available, so that one really does not know if one model's *AIC* is significantly, *in a statistical sense*, different from the criterion for another model. Thus the element of prejudice once again raises its head. If one prefers one model to another despite the "perverse" relationship of the respective *AIC*'s, then one writes something like "the two models *AIC*'s are almost the same and thus I prefer my favorite model to the other nasty one."

As we are dealing with financial data, we have a third method available: perhaps the best model should be the one that can make us the most money. Using the estimated model, one may then forecast the returns on an investment that is based upon the model results; alternatively, one may calculate the mean absolute and mean square deviations of the model forecast and of the actual returns. The best model would then be the one with the smallest forecast errors. The forecasts are based on the last estimated *beta* — and *alpha* where relevant — values. For the *RCF* model, this means using the "steady state" *beta* whereas all other models produce a *beta* with perhaps a modeled deviation from this value.

This method of comparison is not without its problems. First of all, we have chosen to do out of sample forecasting and this presents an almost insurmountable problem: how do we forecast the return to the market index? My solution is to cheat: use actual values as they are available when I make the forecast even if they would not be available to a real investor. Any forecast of the market rate I use would have to be the same for all models and thus not affect the ranking of the different alternatives. Thus I use the actual observations.

The second problem is one of the length of the forecast: by choosing different horizons one may well get different answers. I have chosen to forecast the six months July to December, 1991. If the company is not registered on the exchange for in December 1991, then I use the last available six month period for the forecast. To the extent that different processes determine the risk of holding a company's stock during its last months on the market, these forecasts will perhaps have little value. They were calculated for the sake of completeness.

A final problem is how to proceed when the forecasts are good but the statistics are bad. As this is an academic thesis, goodness of fit statistics cannot be ignored. We should, however, be aware that because a model gives a satisfactory description of historical phenomena does not — except perhaps in Orwell's novel — mean that it will perform as well on the future. We are really caught between two opposing forces: those who demand that a model fit the data and those who require relevance for the real world.

Again, my solution is to hedge: if the statistics are "acceptable" for the model with the best forecast record, then that model will be crowned best for that particular stock. It remains to define what "acceptable" statistics implies.

5.6. Diagnostic Tests

The estimated models have been tested for model adequacy using a number of tests, most of which have already been described in the previous chapter. These include the usual goodness of the fit statistics such as R^2, t-tests and the likelihood ratio test. With the R^2, there is here a choice of whether to present the coefficient of determination calculated from the one step ahead residuals or those from the updated states; both are presented. If one wishes to compare with the OLS R^2, then the later should be used; if one wishes to judge the accuracy of the one-step ahead predictions, then the former should be used. I prefer to use the R^2 from the updated states as the "prediction" is not really a prediction that could be used to make money. When the observation on the index variable is available, the true observation on the variable is also known as it forms part of the index. Thus I would rather have a ready comparison with the OLS estimates. I have also used the OLS in comparison when calculating the likelihood ratio test.

Another common diagnostic is the Ljung–Box statistic to check whether there remains any autocorrelation in the residuals. A significant value for this statistic means that there is some autocorrelation in the residuals; it could also imply that the model is poor. The modified von Neumann statistic also provides a specification check as well as providing a test for first order autocorrelation in the residuals. Still another test is the "recursive t-test". If the model is correctly specified, the statistic τ follows a Student's t-distribution with $T - k - 1$ degrees of freedom. (Harvey (1990), p 157.)

$$\tau = \left(\frac{T - k - 1}{T - k}\right)^{0.5} \frac{\sum_{t=k+1}^{T} v_t}{\left(\sum_{t=k+1}^{T}(v_t - \bar{v})^2\right)^{0.5}} \tag{5.17}$$

Here $\bar{v}$ is the mean of the recursive residuals, v_t.

Three tests for heteroscedasticity are also given. These are, first of all, a simple Goldfeld–Quandt which compares the residual variance in the first third of the sample with that of the last third. The test should be two-sided as the variance may decrease as well as increase. A second test is the $ARCH$ test while the final one is the $cusumsq$ test that was also used in pretesting. As explained above, however, that the standard errors are calculated using equation (5.15) so that the rejection of the null hypothesis of homoscedasticity is not all that troubling.

There arises the question as how to treat models with a structural break. If this break occurs in the last part of the sample, there is not much one can do about it. Estimating a Kalman filter with fewer than the fifty data points that Box-Jenkins take as the minimum for fitting *ARMA* models is not advisable. However, some series show a distinct break in or around 1983. This was during a time of unprecedented growth in the Stockholm exchange due to the deregulation of the Swedish credit market.[9]

Those stocks with a break in the period stretching from 1982 - 1985 could be split into two series and estimated as such. However, this exercise is not performed here as this would expand the number of pages but add little new information. Such estimates have been presented for a few stocks in Wells (1994). The plots of the *beta* coefficients will show a break at the point where the series is split. All results here are based on the entire period.

Hopefully, the model with the smallest prediction errors will also give significant estimates for those statistics that one wishes to be significant — *i.e.*, the *t*–tests and the likelihood ratio — and insignificant ones for the diagnostics. In presenting the models, I have not accepted the model with the best forecast if its diagnostics were poor. After all, social sciences are not exact ones and the researcher's own preferences often color his results. I am no different, expect perhaps in admitting that I am being "unscientific".

One further point, however, should be mentioned as it effects both the likelihood ratio test and the *t*–tests for the variance parameters. The *t*–test is symmetrical, but as variances are positive or zero, the test must be modified to account for the non-negativity restriction. The problem arises from the fact that the statistic studied has a different distribution under the null hypothesis than under the alternative. In fact, the statistic is non stochastic under the null: that is to say, it is zero under the null.[10] Gouriéroux *et al.* (1982) show that the null distribution of the parameters is a combination of χ^2 statistics with the degrees of freedom being dependent upon the number of zero-constraints being simultaneously tested. For example, to test if one of the constrained parameters — and only that one

---

[9]Socialists are funny creatures. Those in Sweden have traditionally been for big business — big business, more jobs, higher pay was the philosophy — and, more true to style, for the regulation of everything down to the number of pieces of bread that one should eat each day (8 was the magic number, but the pun is not there in Swedish). Yet, to their credit, they began the process of the deregulation of the credit system in the early 1980's and even voted through large tax cuts at the end of the decade (which they have since relented). Given these new freedoms and the extra income, many people took out second mortgages and played the stock market. Some made a killing. These were called rather derogatorily "financial puppies" by one of the leading socialists (the chairman of the industrial workers' union). This led to the introduction a sales tax on transactions on the stock market which heralded the beginning of the end of the deregulation boom.

[10]Naturally, similar reasoning applies to the elements of the system matrix, Φ.

— is zero, then the distribution of the test statistic — which is calculated as the square of the usual t-statistic — is $\frac{1}{2}\chi^2(0) + \frac{1}{2}\chi^2(1)$ where $\chi^2(0)$ is a truncated *chi–squared* distribution with unit mass at the origin and $\chi^2(1)$ is a *chi–squared* statistic with one degree of freedom. The critical region, c, may be found by noting that the probability that the statistic under the null hypothesis is greater than c is twice the significance level. That is, if one is testing at the 5% level, then the appropriate critical area is found from the 10% level. It follows that if the square of the t-statistic is greater than c, then the null hypothesis is rejected at the stated level. For the 5% level, the lower bound on the critical area should be 2.71. This compares to the tabulated 5% level of about 3.84: using the standard methods leads to non-rejection in some cases which call for rejection.

When one tests if there are two parameters that simultaneously take the value zero, then the critical area is a bit more complicated to calculate. Gouriéroux *et al.* use a geometrical argument to derive this area. They show that the solution to the following equation yields the relevant critical area:

$$q + 0.5\Psi_1(c) + (0.5 - q)\,\Psi_2(c) = 1 - \alpha \qquad (5.18)$$

Here Ψ_i is the density function for the $\chi^2(i)$ distribution — that is to say, the area under the $\chi^2(i)$ frequency function from zero to c. The constraint $1 - a \geq q$ must hold. While Gouriéroux *et al.* provide a small table, there is really no problem involved extending it; a few lines in GAUSS will suffice. In the case where the alternative hypothesis are of the form $\gamma_i \geq 0$, (5.18) simplifies to

$$\frac{1}{2^p}\sum_{i=1}^{p}\binom{p}{i}\,\Psi_i(c) = 1 - \alpha \qquad (5.19)$$

For the 5% level, $c = 4.2307$ when testing if two independent parameters simultaneously are zero — that is, are simultaneously on the boundary of the parameter space. The corresponding figure for three is $c = 5.4346$ and for four is $c = 6.4976$. If the parameters are not independent, the critical value c is smaller; still the above values are quite a bit less than the tabulated values for the *chi–squared* statistic with 2, 3 and 4 degrees of freedom (5.9915, 7.8147 and 9.4877). The likelihood ratios are compared to the critical values calculated from (5.18) in the case with two parameters that are restricted to be positive. When more than two are restricted, the likelihood ratio will be compared to the critical value calculated from (5.19). Even if this latter comparison will result in failure to reject the null when in fact it would have been rejected had the correlation between the estimated parameters been taken into account, it does present a more correct test that if we had ignored the inequality restrictions entirely.

5.7. Some Problems to be Considered

Before presenting the actual estimates — the subject of Section 5.8 — we should address a few remaining problems. One of these is the question of outliers. Financial data is rather volatile and observations greater that four standard deviations from the mean are by no means unusual. What does one do with such an observation? The second question is that of the likelihood function. It is very flat, and often there is little variation in this function for small parameter perturbations from the maximum solution. While causing problems for the parameter estimations, the flatness of the function seems not to have much effect on the estimate of the state variables which remain much the same for small parameter variations. There is also the question of multicollinearity which especially appears in the estimates where the state variable is assumed to follow a first order autoregressive pattern. This complicates the estimation — especially the Hessian matrix becomes very close to being singular and the resulting standard errors are very small. Below we shall examine each of these three issues.

5.7.1. OUTLIERS

The market model assumes that there is but one random process at work. This is the part of the variation in the returns of an asset that may be reduced by diversification. However, this random process may well consist of two — or more — processes one of which may be very strange indeed. Consider the model

$$y_t = f(z_t) + \varepsilon_{1t} + \varepsilon_{2t} \tag{5.20}$$

Here f is some function transforming the exogenous variable of the system z_t — say the returns to the market index — to the endogenous variable y_t — the returns to the asset at hand. The two other right-hand variables are the disturbances — random variables with different distributions. We might think of both as being normally distributed but with different parameters; or we might consider ε_{1t} to be normal and ε_{2t} to follow another distribution. Equation (5.1) above assumes that $\varepsilon_t = \varepsilon_{1t} + \varepsilon_{2t}$ with ε_t being the non-systematic part of the risk in holding the asset. However, we might just as well let ε_{1t} be the non-systematic risk and assume that the other process — ε_{2t} — is too random to be modeled. It might have an extremely large, or perhaps infinite variance; it might be deterministic except that it's exact appearance in known only to the insiders who can exploit it. However, it is questionable whether this process should be part of the market model.

One way around the problem is to truncate all observations on the returns to the asset that are larger than say three standard deviations away for the mean return. The philosophy behind such a procedure would be that, as the market model is not able to cope with the disturbances ε_{2t}, we should simply try to remove their influence on the model. We identify these observations as those that lie further than 3 standard deviations from the mean of y_t. These are then set equal to the mean plus — or minus if the deviation is negative — 3 standard deviations. This is a rather *ad hoc* approach to the problem. It is quick and somewhat dirty, but there is a test of its relevance: if the prediction errors for the out of sample forecasts are lowered, then this truncation of the dependent variable would seem justified. If the forecast errors are not reduced, then there is no justification for the truncation.

There are of course other methods available. One such approach would be to truncate *residuals* that are too large or small, thereby decreasing the effect of outliers. This method in an *OLS* context is outlined in Judge *et al.* (1985). "Too large" or "too small" is defined in reference to the standard deviation of the residuals from the regression of the independent variables on the dependent. However, the residuals should be those from a robust estimation. An example of a robust estimation would be to replace the objective function in the ordinary least squares regression — that is, the sum of the squared residuals — by one that places lesser weight on large deviations. The sum of the residuals defined as the absolute deviation of the actual value of the dependent variable from its predicted one would be a suitable objective function. A robust measurement of the dispersion of these residuals would be the median of their absolute values. This median could then be used in truncating the residuals in the Kalman filter: all those that we greater than, say, twice this median value would be set equal to the median. Of course, the sign of the residual would remain unchanged. Thus we could obtain estimates of the systematic risk for an asset from the state estimates generated by the Kalman filter. However, the diagnostic statistics are no longer useful. Again, we could test the relevance of the estimation by using out of sample forecasting as suggested above.

It has been my experience that while the states estimated from the truncated dependent variable approach closely resemble those estimated freely, those estimated from the truncated residuals are usually different.

Both approaches are rather arbitrary although the latter one has some support in the literature at least in the *OLS* context. Both use exogenous criteria to adjust the data. The first one is based on the intuition that large deviations cannot be explained by the model, but may also be thought of as a type of "trimmed regression" where the offending variables are truncated rather than discarded. The second approach uses a more "scientific" method but in the end, the choice of the multiple for the median is arbitrary. (see Judge *et al.* (1985), pp. 829-839) My preference is for the first method although both have been attempted. There is one other dis-

Log likelihood for BGB

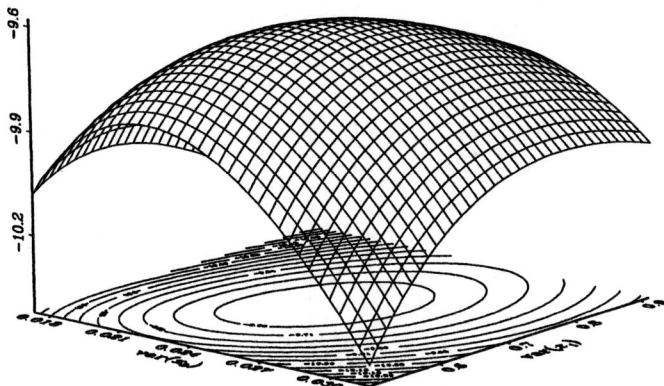

Figure 5.1. **The likelihood function for BGB.** The function is very flat but does exhibit a definite maximum point. The surface is generated for the maximum likelihood estimate for the random walk model for BGB that is presented in Table 5.5. Note that the vertical scale is somewhat miss-labeled: the numbers on this axis must be divided by ten and added to -213.0 to find the value of the calculated log likelihood.

advantage to working with the truncated data: the interpretation of the *cusumsq* plots is not as straight forward if outliers are cut down to size. Even one extreme observation, and the resulting large residual, will often force the *cusumsq* line out of its confidence limits as much of the variation in the entire series will be embodied in this one residual. If the observation were a true outlier, then truncation — by smoothing the *cusumsq* plot — removes a spurious indicator of structural change. If, however, the observation actually heralds a structural break, our data mining has effectively removed the usefulness of this plot in identifying such changes. The data is expressed in rates of return which may well settle down to more "normal" levels once the storm of the initial change has passed. Therefore, the basic results that are reported are those derived from the actual data rather than the truncated data.

5.7.2. THE INITIAL VALUES

The Figure 5.1 will serve as an example of the likelihood functions for the estimates. The point made here is that although the function is rather flat it does seem to have a single maximum. Different starting values all converge — more or less — to this point.[11] This is perhaps the proper place to discuss such values.

There are two different types of initial values that need to be set. One of these is the initial value for the state vector in the Kalman filter as well as its covariance. For mean reverting models as well as the random coefficients model, the initial state — or states as the case might be — are set to zero and the initial covariance is set to a large number in accordance with Harvey's recommendations. For random walk models — I include the ROS model and the moving mean model as well as the random walk model — the initial states are set equal to the OLS estimates obtained from the first 10 observations; the initial covariance of the states is the covariance matrix of these OLS estimates.

The second set of initial values are the ones for the hyperparameters to be estimated by maximizing the likelihood function. Once again, the means in the mean reverting model as well as the RCF and ROS models are easy to set: one simply uses the OLS estimates from the entire sample. The variance of the output equation is for the most part concentrated out of the likelihood function and thus causes no problem. The exception is the ROS model where we set the initial estimate equal to the OLS residual variance.

The real problems are caused by the transition equation. The coefficients in the transition matrix Φ, when it is not zero or an identity matrix, are set to 0.50 for want of a better value: experimentation showed that the final estimates were not at all sensitive to this value. Initial estimates of the variances in this equation are also difficult to justify. For want of better suggestions, I have followed Kool & Bomhoff (1983) in calculating these from the first few observations. First the median of the absolute first differences of the first 10 observations is calculated. Secondly, the median of the absolute value of the deviations of these first 10 observations from their mean is found. The initial variance is then set to the square of the minimum of these two medians divided by 0.6745.[12]

5.8. The "best" estimates

Table 5.1 presents the results in brief for all of the 57 stocks in the data base. Here we have counted the times that each model was "best" according to the five different criteria. The lower the AIC the better the model. Similarly,

---

[11]I might also point out the importance of scaling the data: problems with numerical optimization occur especially when the Hessian matrix is poorly conditioned. That is to say, when the diagonal elements of this matrix vary by more than say a power of ten. Scaling the data will help one obtain a well conditioned Hessian and better estimates.

[12]Kool & Bomhoff claim that this gives an approximation to something we know nothing about. (See Kool & Bomhoff (1983), p. 239)

TABLE 5.1. Best models of the individual
stocks by the different criteria

	MM	MRV	RCF	ROS	RW	OLS
AIC	7	11	14	19	4	2
MAD	16	5	15	3	11	7
$RMSE$	15	8	12	3	10	9
$R^2_{t|t-1}$	1	3	44	1	8	0
$R^2_{t|t}$	15	20	12	2	8	0
chosen	23	17	11	1	3	2

both mean absolute deviation, MAD, of the estimated value from the true
and the square root of the mean square error of the estimate, $RMSE$, are
lower for better models. The two R^2 measures are higher for better models.
The last row in the table indicates the frequency for which each model was
considered best. This last row in the table is somewhat arbitrary as I
required the model to be statistically significant — as measured by the
likelihood ratio — for it to be counted as the best.[13] If the model was
not significant, then the next best and so on was counted as the best. In
general, the RCF model which always estimates significantly is the winner
when the best model was not significant.

Ranking the number of times that the different models within each
criterion are best, and then taking the average of these ranks, we conclude
that the random coefficients model is best overall. It is followed by the
moving mean model, then both the mean reverting and the random walk
— they are tied for third place — and finally the Ohlson-Rosenberg model.
Bringing up the rear is the ordinary least squares estimation of the market
equation. Indeed, had the OLS proved better on the average than the more
complicated models one could well question the relevancy of the rather
complicated estimation procedure followed here.

As indicated above, presenting tables with the best estimates involves
a somewhat arbitrary choice. A reasonable requirement is that the variance of the process that drives the *beta* coefficient be significantly different
from zero. With the exception of the random coefficients model where this
requirement is not fulfilled some of the time, the estimates presented in
Tables 5.2 to 5.8 on the following pages appear to me to be the best. How-

---

[13]This is not quite the whole truth. There were times when the model was significant at
about the 10% level but was quite superior where the criterion in question was concerned.
The number of such exceptions vary from criterion to criterion. For AIC there were 3
such exceptions, for the MAD 8, for the $RMSE$ 9 for the $R^2_{t|t-1}$ none and for the $R^2_{t|t}$ 14.

TABLE 5.2. **Moving mean models**

	Aga	Asea	Garphyttan
$\alpha/10$	$8.14E-2$	$4.00E-20$	$4.15E-1$
t-value	$2.24E+2$	$2.25E+2$	$-1.57E+2$
Φ_{22}	$7.69E-11$	$3.30E-9$	$3.05E-2$
t-value	$2.13E-3$	$9.41E-2$	$6.14E+2$
$\text{var}(\xi_{2t})$	$2.33E-2$	$3.45E-1$	$2.12E+0$
t-value	$4.18E+0$	$5.44E+0$	$3.39E+0$
$\text{var}(\xi_{3t})$	$9.22E-3$	$2.02E-3$	$5.5E-16$
t-value	$9.99E+0$	$1.91E+1$	$2.56E-2$
$\text{var}(\varepsilon_e)$	$2.41E-1$	$2.62E+0$	$7.54E-1$
AIC^a	$2.51E-1$	$2.75E-1$	$7.91E-1$
LR^b	$3.81E+1$	$5.16E+1$	$2.91E+1$
T^c	246	246	246
$R^2_{t|t-1}$ ^{d}	0.438	0.503	0.218
$R^2_{t|t}$ ^{e}	0.437	0.629	0.382
Q^f	$1.96E-2$	$5.00E-3$	$4.2E-11$
$Hetero^g$	$1.06E-1$	$6.21E-2$	$8.12E-6$
$Arch^h$	$1.25E-1$	$2.35E-3$	$2.33E-5$
$v\ Neum^i$	$7.63E-5$	$7.03E-2$	$8.65E-3$
$recur\text{-}t^j$	$7.76E-1$	$9.89E-1$	$5.56E-1$
$cs^{2\,k}$	$2.08E-1$	$6.92E-3$	$5.20E-8$

^{a}Akaike information criterion.
^{b}Likelihood ratio.
^{c}Number of observations.
$^d R^2$ calculated from the recursive residuals.
$^e R^2$ calculated from the updated states.
$^f P$–value for the Ljung–Box test.
$^g P$–value for Goldfeld–Quandt test.
$^h P$–value for the $ARCH$ test.
$^i P$–value for the von Neumann test.
$^j P$–value for the recursive t–test.
$^k P$–value for the $cusumsq$ test.

ever, this is not to say that another observer would make another choice. Indeed, given the number of different rankings shown in Table 5.1, it would be truly amazing if all observers came to the same conclusion.

There is one problem common to many of the estimates: the P-value for the $cusumsq$ statistic — cs^2 in the tables — is often less than 5%. To be more precise, in 10 out of the 21 estimates presented in the tables, this

TABLE 5.3. **Moving mean models**

	Investor	Pharmacia	Åkermans
$\alpha/10$	$2.68E-2$	$6.76E-2$	$1.86E-2$
t-value	$5.88E+1$	$5.12E+1$	$7.35E+0$
Φ_{22}	$2.55E-1$	$1.67E-1$	$1.63E-2$
t-value	$6.40E+0$	$7.07E+0$	$1.41E-1$
$\text{var}(\xi_{2t})$	$7.90E-2$	$6.31E-1$	$3.36E-1$
t-value	$2.77E+0$	$6.28E+0$	$5.84E+0$
$\text{var}(\xi_{3t})$	$5.34E-4$	$2.63E-4$	$8.77E-3$
t-value	$4.69E+1$	$2.35E+1$	$1.31E+1$
$\text{var}(\varepsilon_e)$	$1.93E-1$	$3.60E-1$	$3.22E-1$
AIC^a	$2.02E-1$	$3.79E-1$	$3.39E-1$
LR	$5.04E+1$	$1.72E+1$	$1.05E+2$
T	245	229	233
$R^2_{t|t-1}$	0.653	0.381	0.356
$R^2_{t|t}$	0.646	0.634	0.359
Q	$6.00E-4$	$2.50E-5$	$8.52E-1$
Hetero	$2.22E-5$	$1.50E-2$	$6.11E-1$
Arch	$1.52E-1$	$3.91E-1$	$5.12E-1$
v Neum	$1.17E-2$	$4.63E-1$	$1.81E-1$
recur-t	$6.64E-1$	$8.35E-1$	$8.21E-1$
cs^2	$1.21E-5$	$6.65E-5$	$8.18E-2$

^{a}See Table 5.2 for explanations

statistic gives some indication of structural change.[14] Again, one should not be too surprised that such a large number of stocks seem to have at least one structural break. On the contrary, I am surprised that as many as 11 of them show no such break.

At times the large *cusumsq* statistic indicates an outlier. Garphyttan, for example, in late 1977 has residual that lies more than 4 standard deviations from the mean; Custos, in 1984, and Ericsson in 1989 also have residuals that are as large. Indeed, residuals that are four standard deviations off the mean are not uncommon. Such large residuals may signal the beginning of a new era; they may also be just temporary fluctuation that should be ignored.

Changes in residual volatility also cause the *cusumsq* statistic to peak outside its confidence lines. Here is greater evidence of structural shifts as an increase in uncertainty is not an unreasonable consequence of such

[14]The corresponding figures for the entire sample is 35 out of 55 estimates

TABLE 5.4. **Moving mean model for Munksjö.**

$\Phi_{11}{}^{a}$	$1.40E-1$	AIC	$1.49E-1$
t-value	$9.74E+0$	LR	$4.32E+1$
Φ_{22}	$1.0E-12$	T	232
t-value	$1.45E-3$	$R^2_{t|t-1}$	0.208
$\text{var}(\xi_{1t})$	$9.05E-1$	$R^2_{t|t}$	0.989
t-value	$1.55E+1$	Q	$4.60E-1$
$\text{var}(\xi_{2t})$	$1.99E+0$	$Hetero$	$7.40E-2$
t-value	$8.29E+0$	$Arch$	$7.89E-7$
$\text{var}(\xi_{3t})$	$7.01E-3$	$v\ Neum$	$3.29E-1$
t-value	$7.08E+1$	$recur\text{-}t$	$3.32E-1$
$\text{var}(\varepsilon_e)$	$1.83E-1$	cs^2	$8.78E-6$

^{a}See Table 5.2 for explanations.

a shift. When this increase lay at the beginning of the series, the initial observations were ignored; when it occurred towards the end of the series, then there is little to be done: there are not enough observations to allow the estimation of parameters. The tables present estimates for the entire period rather than splitting the sample for those series that split somewhere in the middle of the period. I present estimates for 21 stocks: it seems to be counter productive to introduce even more estimates here. I will return to this issue in the next chapter.

A second general observation worth commenting on is the estimation of the intercept term in the market model (5.1). In the *OLS* estimation of this equation, the intercept term is almost always insignificant. This would suggest that means could be subtracted for the variable before estimation thereby reducing the number of parameters that need to be estimated. However, the intercept term in (5.1) is also — in some nebulous way — related to the risk free rate of interest. This in turn suggests that the constant should perhaps after all be estimated. Of course, the estimates are much the same whether or not the means are removed; the $R^2_{t|t-1}$ is slightly better with the means removed, but the difference is negligible. I present estimates in where the means and the risk free rate have not been removed.

One further comment should be made before continuing: a true risk free rate of interest available only for the period after March 1981. Until this date, the credit market in Sweden was tightly regulated mainly to keep the rate of interest artificially low in order to keep housing "cheap". During

TABLE 5.5. **Random coefficient models.**

	Custos	Esab	Protorp	Providentia	Trelleborg
$\alpha/10^a$	$-4.68E-3$	$6.26E-1$	$1.77E-2$	$-1.46E-2$	$1.55E-1$
t-value	$-7.41E-1$	$4.94E-1$	$1.37E+1$	$5.07E+1$	$1.84E+1$
beta	$1.13E+0$	$9.32E-1$	$9.58E-1$	$1.33E+0$	$1.05E+0$
t-value	$7.64E+1$	$4.44E+1$	$8.63E+1$	$1.24E+2$	$1.47E+2$
$\text{var}(\xi_t)$	$2.25E-1$	$8.35E-2$	$1.63E-1$	$2.53E-1$	$5.28E-2$
t-value	$3.11E+0$	$2.43E+0$	$5.60E+0$	$6.03E+0$	$5.71E-1$
$\text{var}(\varepsilon_t)$	$2.96E-1$	$5.35E-1$	$3.01E-1$	$1.72E-1$	$7.04E-1$
AIC	$3.05E-1$	$5.67E-1$	$3.21E-1$	$1.78E-1$	$7.42E-1$
LR	$8.15E+0$	$7.09E+0$	$1.84E+1$	$3.16E+1$	$6.51E+0$
T	245	138	125	245	245
$R^2_{t\mid t-1}$	0.490	0.389	0.522	0.699	0.354
$R^2_{t\mid t}$	0.636	0.443	0.665	0.831	0.389
Q	$3.24E-2$	$3.46E-2$	$3.78E-1$	$8.60E-1$	$3.89E-1$
Hetero	$4.00E-4$	$7.85E-1$	$1.35E-1$	$1.42E-2$	$6.50E-1$
Arch	$2.19E-1$	$5.96E-1$	$4.33E-1$	$1.54E-1$	$9.45E-1$
v Neum	$1.09E-1$	$1.09E-1$	$5.60E-2$	$3.72E-2$	$1.03E-1$
recur-t	$7.03E-1$	$8.33E-1$	$1.55E-1$	$8.38E-1$	$1.60E-1$
cs^2	$1.54E-6$	$1.95E-1$	$6.70E-2$	$2.02E-2$	$3.97E-1$

[a]See Table 5.2 for explanations.

the 1970's, the only available risk free rate is for *a vista* deposits in the commercial banks. The rate was low, but funds were available after 30 days (or of course less). In March 1981 treasury bonds with a turn around time of 30 days were made available. Thus this rate has been used as an approximation of the risk free rate in the 1980's and 1990's while the deposit rate was used in the 1970's.

A third point worth noting is the percentage of the different models presented. The "best" model for the individual stocks is the moving mean model. I find this rather surprising. My own prejudice is for the random walk model. One finds references to this specification as early as 1971. The behavioral assumption is that the best prediction of the systematic risk of a stock is the current measurement of that risk. People actually pay money for such estimates and use them in determining investment strategies. Further, the random walk model contains but two estimated parameters and is the easy to estimate, even on a rather primitive computer. When this project was begun, I used a computer based on Intel's 286-chip. It was slow work and estimation time for models with many parameters

TABLE 5.6. **Mean reverting models.**

	Alfa Laval	Ericsson	Pharos	SKF*[a]	Volvo*
Φ_{11}[b]	$9.14E-1$	$8.99E-1$	$4.32E-1$		
t-value	$2.03E+1$	$1.51E+2$	$3.85E+1$		
Φ_{22}	$2.12E-1$	$2.96E-1$	$9.33E-1$	$4.8E-10$	$6.35E-7$
t-value	$1.33E-1$	$1.29E+1$	$8.74E+0$	$9.32E-3$	$4.84E-1$
$\alpha/10$	$7.32E-2$	$3.80E-2$	$1.01E+0$	$-1.58E-2$	$-3.35E-2$
t-value	$8.34E-1$	$2.73E+1$	$3.88E+1$	$2.12E+1$	$3.04E+1$
beta	$1.08E+0$	$9.54E-1$	$4.74E-1$	$1.27E+0$	$1.27E+0$
t-value	$4.60E+2$	$1.71E+2$	$3.97E+2$	$1.05E+2$	$4.53E+2$
var(ξ_{1t})	$4.16E-3$	$1.31E-2$	$1.37E-1$		
t-value	$7.39E+0$	$2.94E+1$	$3.70E+1$		
var(ξ_{2t})	$4.54E-1$	$5.25E-1$	$1.08E-1$	$2.52E-1$	$2.17E-1$
t-value	$6.83E+0$	$7.15E+0$	$3.66E+0$	$5.45E+0$	$1.32E+1$
var(ε_t)	$1.99E-1$	$3.57E-1$	$3.46E-1$	$3.34E-1$	$2.62E-1$
AIC	$2.13E-1$	$3.81E-1$	$3.94E-1$	$3.47E-1$	$2.73E-1$
LR	$2.65E+1$	$2.47E+1$	$5.16E+1$	$9.03E+0$	$5.85E+0$
T	241	246	125	246	246
$R^2_{t|t-1}$	0.494	0.341	0.174	0.633	0.678
$R^2_{t|t}$	0.802	0.717	0.643	0.698	0.734
Q	$3.73E-1$	$8.67E-2$	$1.18E-1$	$1.00E-4$	$3.27E-1$
Hetero	$1.36E-1$	$1.72E-5$	$2.08E-1$	$1.30E-6$	$3.17E-1$
Arch	$6.10E-1$	$5.93E-1$	$6.50E-1$	$1.70E-1$	$9.42E-1$
v Neum	$3.53E-2$	$3.63E-1$	$5.10E-1$	$1.75E-4$	$7.00E-1$
recur-t	$9.31E-1$	$8.81E-1$	$8.39E-1$	$8.75E-1$	$8.58E-1$
cs^2	$4.69E-1$	$6.13E-7$	$3.31E-1$	$8.94E-6$	$2.64E-1$

[a]For stocks marked with an asterisk "*", the model has been estimated using a constant α.
[b]See Table 5.2 for explanations.

took literally hours to converge. I came to love parsimonious models such as the random walk or the random coefficients (with the means and the risk free rate removed before estimation). However, thanks mainly to series of grants from the Tore Browaldh foundation, I was able to purchase machines with much greater capacity: my current one runs an Intel DX2-66 chip and requires minutes for estimating larger models.[15]

[15]I have also a 90 MHz Pentium with the now infamous floating point bug: thus I still do my estimations on the older machine in spite of the potential time gains. The Pentium runs the same program in about 25% of the time required on the DX2-66 machine. Intel has, however, promised to replace the faulty chip.

TABLE 5.7. **Random Walk models.**

	BGB	Gambro	Industri-värden[a]
$\text{var}(\xi_2 t)$[b]	$7.88E-3$	$3.11E-3$	$2.10E-5$
t-value	$1.37E+1$	$3.46E+1$	$2.55E+1$
$\text{var}(\xi_2 t)$	$3.33E-2$	$6.11E-3$	$3.15E-3$
t-value	$3.53E+0$	$5\text{-}07E+0$	$7.98E+0$
$\text{var}(\varepsilon_t)$	$6.55E-1$	$6.93E-1$	$2.50E-3$
AIC	$7.11E-1$	$7.52E-1$	$2.55E-3$
LR	$6.31E+0$	$5.04E+0$	$1.80E+1$
T	98	98	245
$R^2_{t\|t-1}$	0.426	0.403	0.579
$R^2_{t\|t}$	0.569	0.455	0.606
Q	$4.98E-1$	$2.62E-1$	$5.40E-3$
Hetero	$7.69E-2$	$2.94E-2$	$1.20E-1$
Arch	$3.60E-1$	$8.83E-1$	$9.55E-1$
v Neum	$1.75E-1$	$5.81E-1$	$3.29E-1$
recur-t	$8.27E-2$	$1.48E-1$	$1.15E-1$
cs^2	$5.69E-1$	$4.31E-3$	$6.35E-2$

[a]For the data for Industrivärden was divided by 100 rather than by 10 before estimation.

[b]See Table 5.2 for explanations.

Then an anonymous referee for *Applied Financial Economics* suggested that I use a model with a moving mean for *beta*; he referred to a paper by Lee & Chen (1982). In this article, *beta* was assumed to be time–varying according to

$$\beta_t = \bar{\beta} + \gamma_1 t + \gamma_2 t^2 + \eta_t \tag{5.21}$$

Here t is a time trend, the η_t is a random component and the bar above *beta* indicates that this part of the equation is the "constant component of the *beta* coefficient" (Lee & Chen (1982), p. 202). I object to this formulation basically because of the quadratic terms; however, as it stands, the equation does not explain the mean of the coefficient. It explains the coefficient itself.

One could of course ask why the mean of the risk coefficient should not be constant. I have two answers. The first one is to return the question with a "why not" added. The risk of holding a financial asset is related to the type of asset one holds and to the prospects this firm faces. This

TABLE 5.8. Rosenberg–Ohlson model
for Export.

$\Phi_{11}{}^a$	$2.26E-1$	AIC	$2.09E-1$
$t\text{-}value$	$3.68E+1$	LR	$2.71E+1$
$\alpha/10$	$5.80E-3$	T	243
$t\text{-}value$	$4.68E+0$	$R^2_{t|t-1}$	0.448
$gamma$	$9.04E-1$	$R^2_{t|t}$	0.734
$t\text{-}value$	$3.03E+1$	Q	$3.09E-6$
$lambda$	$2.52E-3$	$Hetero$	$9.05E-2$
$t\text{-}value$	$2.13E-1$	$Arch$	$4.48E-2$
$var(\varepsilon_t)$	$1.94E-1$	$v\ Neum$	$7.92E-7$
$t\text{-}value$	$2.59E+2$	$recur\text{-}t$	$7.76E-1$
$var(\xi_t)$	$3.31E-1$	cs^2	$6.50E-2$
$t\text{-}value$	$1.25E+1$		

^{a}See Table 5.2 for explanations.

is, however, not the entire picture. Holders have their own ideas as to future market growth and profitability. Their actions — which of course are independent of the firm whose asset they are considering — also contribute to the variability of the risk picture. The idea behind a moving mean is that at each point in time there exists an unobservable *beta* value. The coefficient that we measure is a combination of this unobservable variable and other purely random occurrences. I call the unobservable component the "mean" of the coefficient.

My interpretation of *moving mean* models is rather far removed from (5.21); however, my equation (5.7) encompasses many different models. Such flexibility was indeed reason that Lee and Chan considered their equation. If we examine my model, we discover just how flexible (5.7) is.

If $var(\xi_{1t})$ estimates to zero, then the constant in the market equation is found to be just that: a constant. The model may then be respecified so that this constant is included in the observation equation (5.1) and the dimensionality of the state vector reduced by one. Of course there are still four hyperparameters to be estimated. In Tables 5.2 – 5.3, all of the models were estimated with a constant intercept rather than one that followed a random walk. The exception is Munksjö in Table 5.4.

The third state variable in (5.7) is the mean of *beta* which is assumed to develop as a random walk. Without the random variations around this variable — that is, if the variance of $var(\xi_{2t})$ were zero and Φ_{22} were unity— then (5.7) is identical to the random walk model. Thus what was said in justification for the specification of the mean of *beta* developing as a random walk is also valid for the random walk model itself. If Φ_{22} as well as $\text{var}(\xi_{1t})$ and $\text{var}(\xi_{3t})$ estimate to zero, then the model becomes the model becomes the random coefficients model: *beta* varies randomly about a constant value which one can interpret as the "true" or "long run" value of the coefficient. This model is of course almost the same as the *OLS* estimate of the market model. However, it has greater flexibility in that *beta* is allowed to vary about its mean value. The main problem with this model is that — given a small *P*–value for the *cusumsq* statistic — is that the mean often shifts at some point during the period studied. There is of course nothing wrong with discrete shifts in means; indeed, by the fact that the sample is split and two — or more — equations estimated one assumes a discrete shift in a hyperparameter has occurred. However, given the number of different agents involved in the market, it would be surprising if everyone changed opinions at once. It seems more logical that the shift would be gradual. This is philosophy behind the moving mean model.

The third model that may be obtained from the moving mean one is the mean reverting model. In the literature on financial series, "mean reverting" has a slightly different connotation than that I use here. There the concept refers to the autocorrelation function of a series. Thus a series is found to be mean reverting if the first term in this function is negative: not only does the series revert to a lower — and unspecified — value, it forever jumps past it being too high one period and too low the next. The behavioral assumption is that the agents always over-react. I find little of merit in this position: if one wishes to discuss mean reversion using an autocorrelation function, then all that is needed is for the function to have roots inside the unit circle. Such a function exhibits a stable rather than an explosive development over time: that is, the effects of one exogenous impulse decrease the further one comes for the time of the chock. In the mean reverting model that I present, the estimated coefficients tend to return to their estimated mean. The transition matrix is diagonal where the non-zero elements are constrained to be positive. Negative elements introduce the "over-shooting" behavior discussed above.[16] Positive elements of course preclude such behavior. One finds both approaches in the literature; some authors restrict to positive values, while other simply restrict to the interval between plus and minus one.

5.9. Reflections on the tables

According to the tables, about one third of the 21 stocks estimate as a moving mean model, 24% as a random coefficients, 24% as a mean reverting

---

[16]See pages 21, 87–88 and 93–94.

and 14% as a random walk model. In general, the diagnostics are satisfactory but there are cases — about 9 — when these statistics are poor. For the most part, the poor diagnostics result from a structural break in the series and were the series split into two or more sub-samples, the diagnostics would improve immensely. The most satisfactory part of the regression is the "large" value for the R^2 statistics, especially the ones calculated from the updated states as these are directly comparable to those obtained from the OLS estimation of the market equation. There are 6 of the 21 greater then 0.7; 8 are between 0.5 and 0.7; and 7 are between 0.3 and 0.5. There is no really low $R^2_{t|t}$ (less than 0.3). This compares to the 0.3 - 0.4 that is the rule rather than the exception in the OLS estimations. While the market model is a simple model of a complex process, the estimations here uphold its relevancy, at least when one estimates with variable coefficients.

As second point worth noting is that for only one of stocks has an estimate that suggests that the OLS model might suffice. This stock is Trelleborg in Table 5.5. The pretests in Table 2.2 were negative for this stock while those in Tables 2.1 and 2.3 suggested variable coefficients. This in turn prompts us to ask whether the pretesting in the case at hand is really necessary. The basic problem here is that asking general questions — are coefficients constant or not — provides general answers while specific questions — are the coefficients constant or do they move as a random walk — yields specific answers.

My preference is for the general type of pretests. A number of alternative model specifications are possible. If the estimates of all but one of the models prove to be statistically insignificant, then there is no problem in choosing a model. However, the estimates are often significant for more than one of the models and one must choose which is best. This is one of the issues addressed in the next chapter.

The bottom line on the estimates presented here seems to be that variable coefficient models are both feasible and informative. The weakness with some of the models is that they seem to span two or more regimes: they should be reestimated for those cases where structural change is indicated.

A final observation concerns the lack of $ARCH$ tendencies in the residuals. Only 4 of the estimated 21 stocks reject the null of no $ARCH$ in the residuals at the 5% level. 3 of these 4 also show evidence of structural change, which, when accounted for, removes the $ARCH$ tendencies. This result seems to be at odds with other findings that $ARCH$ processes in asset returns are common. I suggest that constant coefficient models are an incorrect way to model asset returns. Once the inherent variability in the coefficients is accounted for, much of the $ARCH$ tendencies disappear.

Chapter 6

The estimates, reconsidered

6.1. Introduction

In this chapter we reconsider some of the estimates. The first order of business is to address the question of those stocks for which the diagnostics were poor. We concentrate on 9 of the stocks, but what is said could just as well have been applied to the others with less than satisfactory estimation statistics.

The second question that is approached is the difference between models: if, say, the moving mean model is preferred to another, how would the estimated *beta* coefficient appear had another model been presented. Once again, we discuss the issue using but small number of stocks — five in this case.

6.2. Some reestimations

Tables 6.1 and 6.2 present the "updated" estimations. Using the information in the plot of the *cusumsq* as well as the normalized residuals, the sample was split into periods of more or less equal residual variance. As some of the stocks proved very volatile during that last few months of the sample (late 1990 through the end of 1991), the only measure that could be taken is to shorten the sample. This is, of course, not very satisfactory if one wishes to forecast the future, but given a longer sample, this final part of the period could be estimated with greater accuracy. Thus, for Investor and SKF, the estimates presented are simply "for the record": they indicate that the method is acceptable even if the data indicate some shift in the hyperparameters that cannot be estimated give the current sample length.

Asea reestimates quite satisfactorily as a moving mean model. That the variance of ξ_{2t} is not significantly different from zero implies that the Asea's *beta* is best described as random fluctuations about a mean that

119

TABLE 6.1. Some reestimated models

	Asea	Custos†	Esab	Garphyttan	Investor†
period	83:1-91:12	75:1-91:12	79:1-91:12	83:1-91:12	82:1-91:6
Model	*MM*	*RCF*	*MM*	*MM*	*RCF*
Φ_{22}	$4.50E-1$		$6.81E-1$	$7.33E-1$	
t-value	$8.77E+0$		$9.63E+0$	$8.89E+0$	
$\alpha/10$	$9.00E-2$	$-4.66E-2$	$8.74E-2$	$2.64E-2$	$-6.52E-4$
t-value	$1.06E+0$	$2.98E+00$	$7.99E+0$	$1.10E+1$	$-2.05E-1$
$\text{var}(\xi_{2t})^a$	$4.31E-2$	$1.07E+0$	$5.84E-2$	$1.05E-1$	$1.32E+0$
t-value	$9.93E-1$	$5.55E+1$	$2.86E+0$	$3.51E+0$	$7.46E+1$
$\text{var}(\xi_{3t})^b$	$3.83E-3$	$1.69E+0$	$5.72E-3$	$1.33E+1$	$2.94E-2$
t-value	$6.35E+0$	$1.69E+0$	$1.45E+1$	$3.51E+0$	$6.46E+1$
$\text{var}(\varepsilon_t)$	$3.16E-1$	$3.71E-1$	$5.38E-1$	$4.18E-1$	$2.95E-1$
AIC	$3.55E-1$	$3.89E-1$	$5.86E-1$	$4.18E-1$	$2.95E-1$
LR^c	$3.64E+1$	$5.54E+1$	$3.10E+1$	$5.03E+1$	$3.15E+0$
T	102	168	138	95	114
$R^2_{t|t-1}$	0.568	0.475	0.349	0.276	0.721
$R^2_{t|t}$	0.585	0.594	0.379	0.382	0.745
Q	$2.15E-1$	$1.97E-1$	$7.48E-1$	$6.44E-1$	$4.94E-1$
Hetero	$7.66E-1$	$1.27E-3$	$4.97E-1$	$1.77E-1$	$1.50E-1$
Arch	$8.90E-1$	$3.38E-1$	$4.08E-1$	$2.19E-1$	$8.45E-2$
v Neum	$3.93E-1$	$1.99E-1$	$7.25E-1$	$9.12E-2$	$2.92E-2$
recur-t	$7.19E-1$	$7.57E-1$	$5.62E-1$	$8.70E-1$	$9.93E-1$
cs^2	$7.83E-1$	$9.64E-4$	$1.14E-1$	$1.18E-2$	$1.28E-2$

^{a}For series marked with a "dagger" (†), the table value is the estimate of *beta* rather than var(ξ_{2t}).

^{b}For series marked with a "dagger" (†), the table value is the estimate of the constant *beta* rather than var(ξ_{3t}).

^{c}See Table 5.2 for explanations of the tests.

follows a random walk. If a *RCF* model is estimated, one finds that the *OLS* model has a smaller *AIC* as well as smaller forecast errors. However, the *MM* model out performs both of the these other two on all criteria. For example, the forecasts for both the *mad* — 8.621 — and the *MSQE* —10.605 — are smaller than the corresponding ones for the *OLS* equation — 8.818 for the *mad* and 11.016 for the *MSQE*. Indeed, the R^2 for the one step ahead predictions is much better than the value presented in the previous chapter (0.585 compared to 0.503) although the $R^2_{t|t}$ is somewhat poorer. I should note that these forecasts are not quite as good as those that obtain from using the entire sample.

TABLE 6.2. Some more reestimated models

	Munksjö	Pharmacia	SKF	Trelleborg†
period	83:1-91:12	82:1-90:7	79:1-91:6	83:1-91:12
Model	RCF	RCF	RCF	RW
$\alpha/10$	-1.34E − 2	-4.01E − 2	-9.38E − 3	
t-value	3.01E + 1	-1.06E + 1	-6.58E + 0	
beta[a]	1.40E + 0	1.11E + 0	1.21E + 0	1.34E − 2
t-value	6.18E + 1	3.72E + 1	2.83E + 2	5.42E + 1
var(ξ_t)[b]	1.14E + 0	8.49E − 1	2.31E − 1	5.58E − 3
t-value	5.01E + 0	3.89E + 0	3.46E + 0	1.22E + 1
var(ε_t)	5.87E − 1	3.74E − 1	3.83E − 1	7.80E − 1
AIC	6.42E − 1	4.06E − 1	4.05E − 1	8.39E − 1
LR^c	9.80E + 0	1.18E + 1	9.74E + 0	1.85E + 1
T	89	97	144	102
$R^2_{t|t-1}$	0.336	0.429	0.532	0.430
$R^2_{t|t}$	0.755	0.764	0.679	0.468
Q	8.20E − 1	2.64E − 1	3.82E − 1	7.34E − 1
Hetero	8.44E − 3	1.65E − 1	2.15E − 1	4.82E − 1
Arch	2.61E − 1	2.24E − 1	5.08E − 1	6.07E − 1
v Neum	7.43E − 1	7.99E − 1	4.32E − 2	1.32E − 2
recur-t	8.19E − 1	9.17E − 1	8.69E − 1	1.87E − 1
cs^2	4.53E − 3	9.13E − 1	7.56E − 1	1.23E − 1

[a]For series marked with an "dagger" (†), the table value is var(ξ_{1t}) instead of *beta*.
[b]For series marked with an "dagger" (†), the table value is var(ξ_{2t}) instead of var(ξ_t).
[c]See Table 5.2 for explanations of the tests.

Custos in its reestimated form is perhaps not much better than the model estimated with the entire sample. The estimate presented in Table 6.1 has managed to correct the low Q-statistic and to reduce somewhat the heteroscedasticity, but the price is lower R^2's. Still, the forecasts are a bit better with the shortened sample (10.482 against 10.504 for the mad and 11.876 to 11.904 for the $MSQE$). The variance for Custos is larger during the second half of the sample; further there are two outliers (August 1983 and October 1984). It is these latter two points that force the cs^2 outside it's 5% limits; it is the large variance which lead to lower R^2's.

If there was little to be gained by reestimating Custos, then there is very much to be gained by the reconsideration of Esab: by splitting the series at the start of 1979, one avoids the period May to September 1978 when the returns to Esab were extremely volatile. For example, the stock fell by more than 28% in June only to rise by about 32% in July only to fall by 26% in August. This was the result of a combination of factors including a falling demand for steel (one of Esab's main products) in general and at shipyards in particular, a low demand for welding equipment (another big item for the firm) and even the Algerian nationalization of imports which hit at Esab's import sales company there. In the reestimated model, which is MM rather than RCF, all the diagnostics are well above the 5% level; all the estimated coefficients are significant. Even the mad of the forecast is lower for the OLS estimate being 1.040 for the estimated model against 1.052 for the OLS.

Garphyttan should be split at two different points as its residual variance is largest from late 1977 until late 1982. Here I have present only the latter period. The $beta$ as estimated in the original table is dominated by noise which is especially noticeable in the late 1970's as well as 1982. By estimating a new model for the last part of the sample, most of the diagnostics are acceptable, although large residuals in 1985 still result in a positive test for heteroscedasticity. If one splits at January 1986, then this heteroscedasticity disappears but the model estimates as a random coefficients instead. However, the least squares estimates give the best forecasts. Perhaps more than any other stock discussed here, Garphytt illustrates the problems encountered in modeling returns.

Turning to Investor, I find that the main problem is the increased volatility at the end of the sample; however, removing the last part of 1991 uncovers a shift in 1982 which was more or less dwarfed by the large residuals in later year. Note that Investor now estimates as a random coefficients model. I might also add that while AIC for the OLS model is slightly lower, the forecasts are about 10% better for the RCF model.

Munksjö is another problem. The estimated presented in Table 6.2 is colored by some heteroscedasticity caused by rather large residuals (a bit less than three standard deviations larger than the mean in August and November 1983 as well as in June 1986) which may be removed by beginning the estimation one year later. However, the forecasts are much better using the estimates presented here. Even the $R^2_{t|t-1}$ show some improvement and the $ARCH$ tendencies have been removed.

As with many other stocks in the data base, the hyperparameters for Pharmacia shift in the beginning of 1982. Like Investor and Munksjö, the model for Pharmacia presented in Table 6.1 is a random coefficients model rather than the moving mean one. This new model is much better: not only has the updated R^2 increased from 0.63 to 0.76, but all the diagnostics are now acceptable. The $cusumsq$ indicates structural stability for the period used.

The hyperparameters for SKF show a shift somewhat earlier — 1979 — and, as explained earlier, also in 1991; only the former shift can be accounted for here. This stock was previously estimated as a *MRV* model but is now a *RCF*. Compared with the earlier model, the reestimated model has acceptable diagnostics. Finally, Trelleborg is best described by a *RW* rather than a *RCF* model. The *RCF* is about as good judging from the forecasts and the *AIC* as well as the statistical tests, but the latter seems to be troubled by multicollinearity. The *OLS* model has the lowest *AIC* and perhaps should be preferred.

6.3. The different models

The importance of the model chosen for *beta* cannot be over emphasized: not only do the basic diagnostic statistics and the forecast change as the model changes; even the time path of this coefficient will change as the model representing it is changed. Put in other words, the state representation of *beta* is not invariant to the model chosen. While there is some degree of correlation in *beta* between the models, this depends both upon which stock one chooses and which models are being compared. I have chosen a few stocks to exemplify the issue. The basis for selection has been the likelihood ratio: only those estimates which are significant are compared here.

6.3.1. AGA

Aga's "thing" is industrial gas: it is one of the largest producers in the world and has factories scattered throughout the globe. This is also the first stock in the data base and as such has served as the one that I have "played with" the most. It is a rather good candidate for experimentation as it shows no structural break during the period in spite of changes in policy by the management. It is also one of the less volatile stocks in the data base. Three of the models estimate satisfactorily: the moving mean, the random coefficients and the random walk. Figure 6.1 shows the *beta* estimated using these three different models. The random walk model is interesting: *beta* begins low — the initial estimate of the state is obtained from the estimate from the first 10 observations — and climbs towards a peak at the beginning of 1976. From then onwards, *beta* falls more or less constantly throughout the period. There is no evidence of a single mean value; nor is there evidence of shifting means as the cs^2 statistic lies within its 5% bounds throughout the sample. It is therefore not surprising that the random coefficients model does not tell the same story: here we find random variations that lie above the mean until mid 1975 and then afterwards fluctuate around it. There are, of course, similarities: the sharp

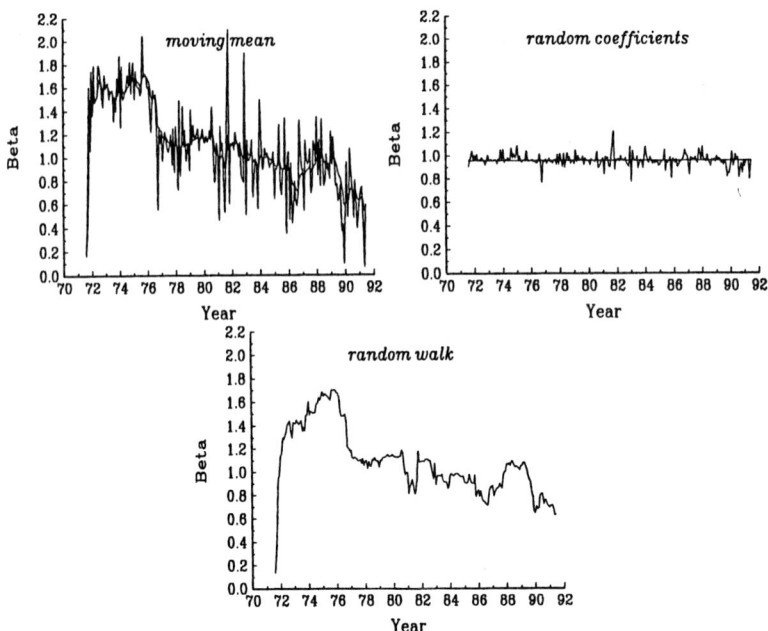

Figure 6.1. The estimated *beta* for AGA. The estimates for three different models. Note the similarity between the mean of *beta* for the moving mean model and the *beta* of the random walk model.

peak in late 1981 corresponds to the upward jump in the random walk *beta* at the same time; the falling tendency in late 1989 is also noticeable in both models. That these two models tell somewhat different stories is revealed by the low correlation between the *beta* of the two models: 0.27.[1]

If one believes in "underlying trends" but doubts if behavior is "smooth" then the moving mean model should appeal: here the "trend" is a random walk about which *beta* fluctuates. The volatility of these fluctuations depend upon two factors. The first is naturally the variance of the process that drives this particular state. The second is the "autocorrelation" coefficient by which I mean the Φ_{22} coefficient of the state equation. The closer this is to unity, the smoother the variations about the mean as the previous state acts as a brake on the system. If Φ_{22} is close to zero, as is the case here, the fluctuations are more or less random. The correlation between this model and the other two is 0.54 and 0.89 respectively for the *RCF* and the *RW* models. Note even though this model has rather random fluctuations, it is more closely correlated with the random walk model than the random coefficients one, as the moving mean model is not constrained to vary about

---

[1]As the estimates of *beta* in the *RW* and *MM* models take a while to initiate, I have calculated all of the correlation coefficients by beginning 13 observations into the series for *beta*.

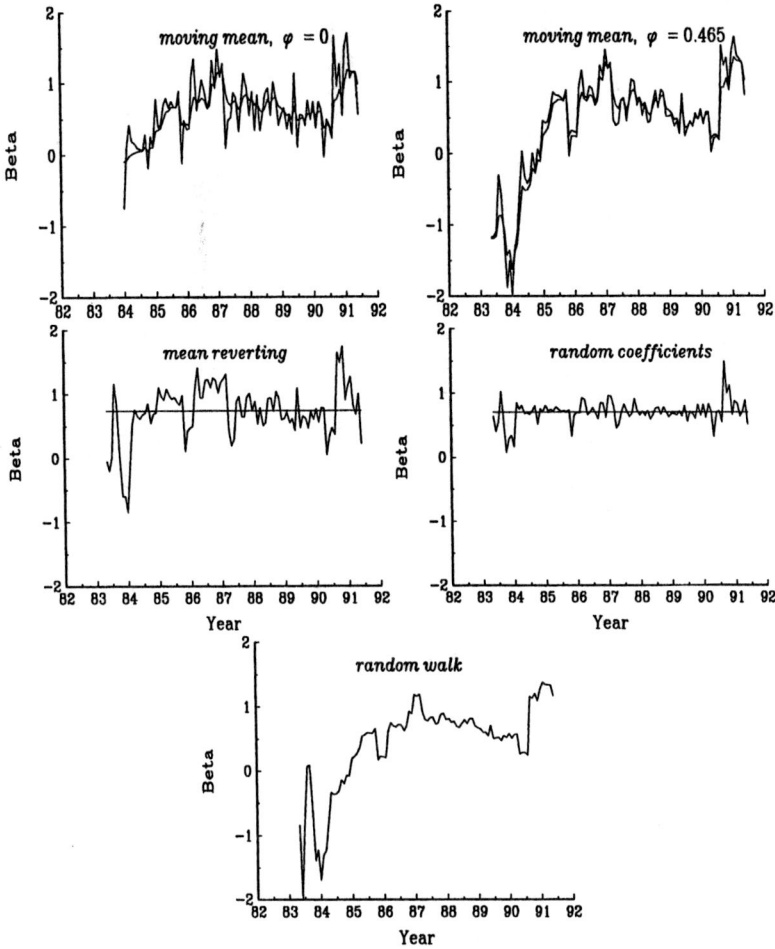

Figure 6.2. The estimated *beta* for BGB. Again we see the similarity between the estimate of the mean of *beta* in the moving mean models and the *beta* in the random walk model. Note also that *beta* is much more volatile when Φ_{22} is constrained to zero than when it is estimated. Note also that the vertical scales are not the

a fixed number. However, the correlation between the moving mean model and the random coefficients one is higher than between the latter and the random walk model as random fluctuations are allowed in the model. Finally, the difference between the mean of the moving mean model and the *beta* of the random walk model is rather small: the correlation between the two is 0.94.

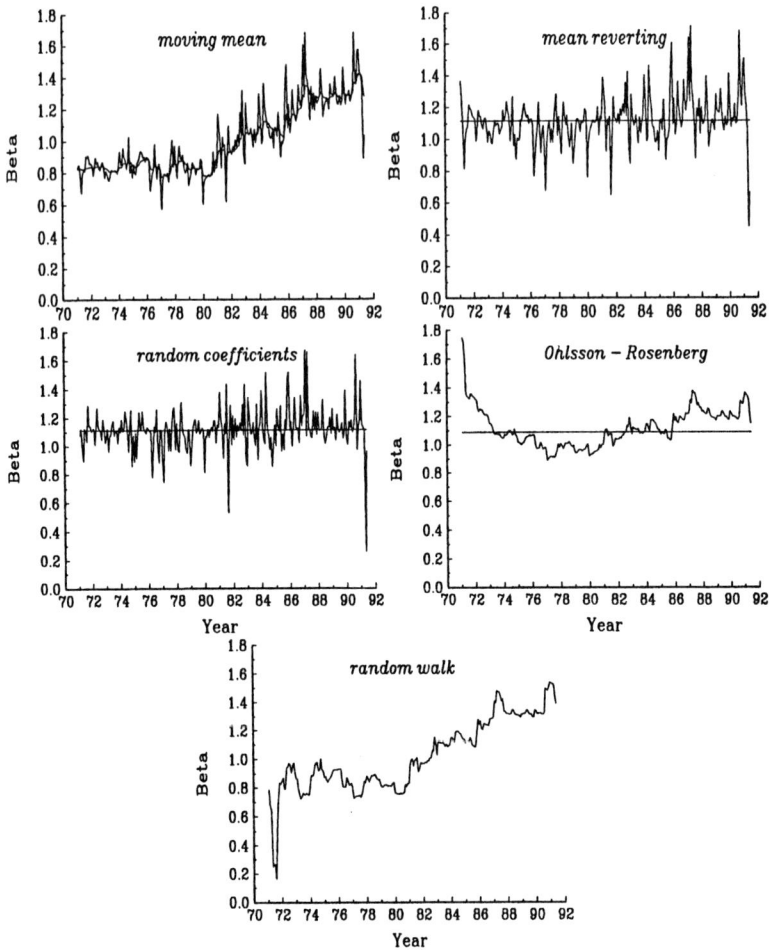

Figure 6.3. The estimated *beta* for Investor. Here one finds a similarity between the mean of *beta* in the moving mean model and *beta* in both the Ohlson-Rosenberg and the random walk models.

6.3.2. BGB

BGB is an investment company specializing in real estate. *beta* for this company begins at a low — indeed, negative — level and climbs to a rather constant value of 0.6 before rising at the end of the period. The various models here tell almost the same story: the exception being the random coefficients model which fluctuates about 0.70 and remains positive even at the beginning of the period. The correlation between the *beta* of this model and the others is not high: 0.34 compared to the random walk model at the lowest and 0.72 compared to the moving mean and 0.77 to the mean reverting at the highest. On the other hand, the correlation between the

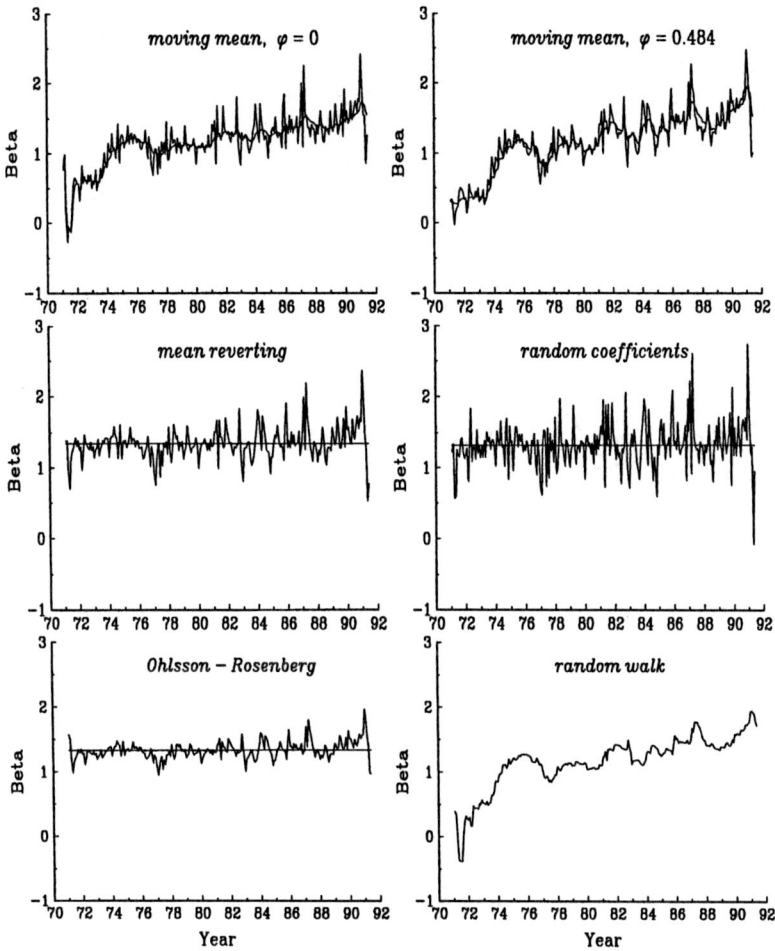

Figure 6.4. The estimated *beta* for Providentia.

random walk and the mean of the moving mean models is high: 0.93 when Φ_{22} is zero and 0.99 when $\Phi_{22} = 0.465$ is estimated. Note the difference between these last named models. When the "auto regressive" coefficient is estimated, the fluctuations around the mean value are dampened: the abrupt jumps observed when Φ_{22} is zero become much smoother. Also noteworthy is the similarity between the mean reverting and the moving mean models. The direction of *beta*'s drift is the same in both models: however, as the former has a mean which is restricted to be constant, *beta* cannot climb to the level obtained in the other model.

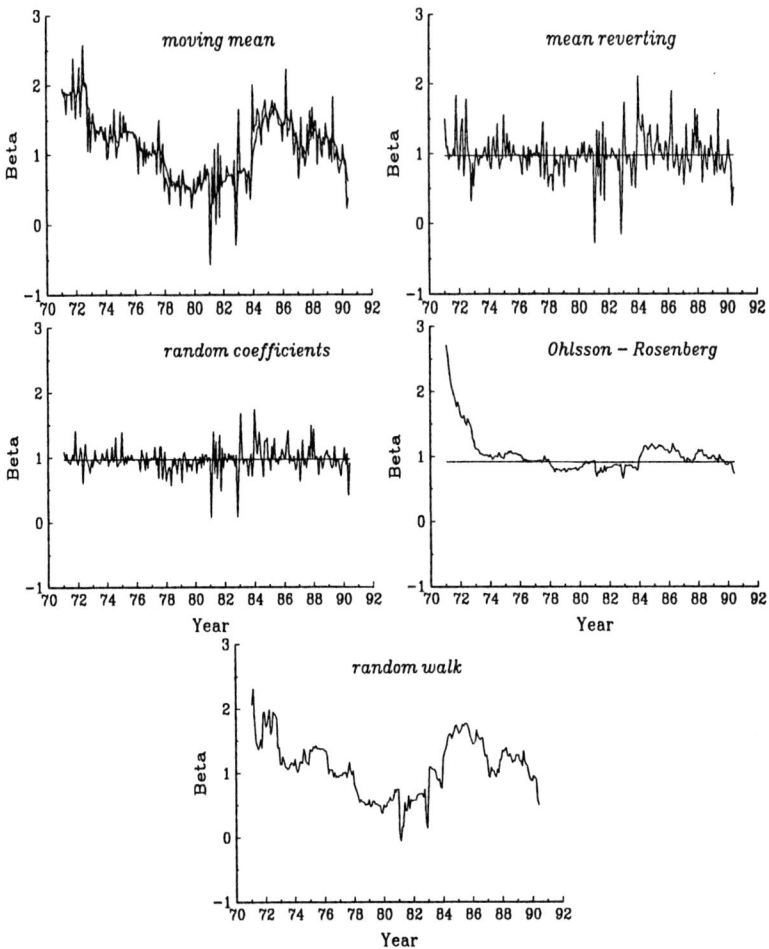

Figure 6.5. The estimated *beta* for Åkermans. Here it seems that *beta* is riding a roller coaster. Its ups and downs may even be detected in the models with a constant mean.

6.3.3. INVESTOR

Investor is a holding company. The *beta* plots for this company include one for the Ohlson-Rosenberg model. Like the mean reverting and the random coefficients models, this one also posits random variations about a "steady state" value for the systematic risk coefficient. However, even if, as above, the Φ_{22} coefficient serves to dampen these deviations from the mean, the correlation between this model and the others is small: 0.54 to the moving mean model, 0.39 to the mean reverting, 0.22 to the random coefficients and 0.41 to the random walk. Here the other two models that vary around a given mean are rather highly correlated (0.86) which is not surprising as Φ_{22} for the mean reverting model estimates close to zero. Again, the highest correlation found is between the mean of the moving mean model and the random walk: 0.94; even the *betas* are highly correlated (0.88) between these two models.

6.3.4. PROVIDENTIA

Like Investor, Providentia is a holding company; and like Investor, its *beta* has a slight upward trend throughout the period, excepting, of course, the mean reverting and the random coefficients models. Unlike Investor, we find here rather high correlation between the models' estimated *beta* coefficients: 0.95 between the Ohlson-Rosenberg and the mean reverting, 0.89 between the random walk and the moving mean with $\Phi_{22} = 0$ and 0.91 when $\Phi_{22} = 0.484$. Even the mean reverting and the random coefficients have a correlation as high as 0.86. Finally, the correlation between the random walk and the mean of the moving mean model is 0.99. Notable also is that the estimated means, when constant, lie very close to each other: 1.34 for the mean reverting, 1.32 for the random coefficients and 1.33 for the Ohlson-Rosenberg models.

6.3.5. ÅKERMAN'S

Åkermans builds construction equipment: its *beta* falls throughout the 1970's and then peaks during the late 1980's only to fall back to approximately the same value with which it began the decade. Correlation of the *beta* coefficients between the models is not as a rule high: only that between the moving mean and the random walk as well as that between the mean reverting and the random coefficients models is in excess of 0.90 — 0.90 for the former two and 0.91 for the latter. That there is a substantial degree of agreement between just these pairs of model is not unexpected. Nor is the 0.95 correlation between the mean of the moving mean and the random walk models.

6.4. The final states

Table 6.3 lists the final state for these various models. Ones immediate impression is that the final *beta* for the different models all seem quite different. If one were to use these final states as a basis for investment strategy then one must choose between risk assessments that present entirely different pictures of the stocks in question. There are of course some obvious choices. One would be to use the model that has forecast the best in the past; however, good historical performance may not be repeated in future forecasts. A second possibility would be to use the one with the smallest *AIC* or residual variance or perhaps the largest likelihood ratio. Still, these

TABLE 6.3. The final *betas*

	MM	MM-a	MRV	RCF	ROS	RW
Aga	0.59	na	na.	0.95	na	0.63
BGB	0.55	0.80	0.23	0.51	na	1.67
Investor	na	1.03	0.67	0.96	1.15	1.39
Providentia	1.25	1.10	0.77	0.94	0.97	1.71
Åkermans	0.36	na	0.51	0.90	0.73	0.50

measures refer to historical and not future development and perhaps are just not relevant. Another choice would be to use an average value — or perhaps the median value. This course of action has the disadvantage the it is not based on the estimated model.

6.5. Lessons

As illustrated in the Section 6.3, state space representations are not unique. The diagrams of the *betas* for some of the models is as well as the table of the final *beta* value should convince the reader that choice of a model must be made with care as the exact information in it depends to some extent upon the model specification.[2]

Which brings me back to my own prejudice. I have philosophical difficulties in accepting a model where an asset's risk coefficient remains constant over a longer period of time. Initially I was drawn to the random walk model. The earliest applications of the Kalman filtering methods to financial data assumed the *beta* developed as a random walk; even latter authors have used this model; tests have been suggested for identifying it. The other model that has received much attention is the random coefficients model: but here one runs into the problem of what the coefficient varies randomly about. As mentioned above, a "steady state" *beta* spanning a 20 year period is just not realistic. One way out is to assume that the steady–state *beta* shifts at discrete intervals that may be identified us-

---

[2]An anonymous for the *Journal of Banking and Finance* once suggested that any verbal description relating the time development of *beta* to the development of the firm as manifested in the news media was irrelevant as different events could be given different interpretations. He suggested relating *beta* as estimated to actual characteristics of the firm such as the debt-equity ratio. Perhaps the reader will appreciate why that this referee was anonymous: as the time path of *beta* varies with the model chosen, any such attempt to correlate firm characteristics with a given estimate will be just as "arbitrary" as he claimed historical descriptions were: he really did not understand the estimation procedure!

ing tests for structural shifts. Another alternative seems to be the moving mean model. Here the mean of *beta* develops as a random walk while the coefficient itself varies around this value. The mean will display a rather smooth time path — see the figures at the end of this chapter — but the coefficient itself will have a lead a more uncertain life. As the correlation between this moving mean and the *beta* of the random walk model is high, it could be studied to reveal the historical development of the firm whose stock is being examined. The only real problem arises when the estimates of this model are poor: if, for example, the variance of the mean in the state equations estimates to a very small number, then the model collapses to the random coefficients or the mean reverting model depending upon the estimate of the autoregressive parameter Φ_{22}. In these cases, my prejudice turns out to be just that: and I must accept the idea of a "steady state" *beta*.

Chapter 7

Modeling with the Kalman filter

7.1. Introduction

Economic literature abounds with fixed coefficient models that could just as easily be estimated with the techniques outlined in chapters 4 and 5. There is no particular reason why variable coefficient models should be the exception rather than the rule. Indeed, that coefficients remain constant over a period of 15-20 years seems more unbelievable than the opposite. However, as the statistical methodology for fixed parameter models is more familiar to economists than those required for estimating variable coefficient models, it is models with constant parameters that abound in the literature. This chapter will present some examples of models that have been presented as ones with constant coefficients that just as easily could have been estimated using the techniques of this book.

I have already mentioned the simple consumption or savings function. Another example would be models that explain the short term rate of interest. A family of such models is presented in Chan *et al.* (1992). A third would be a simple example of an *ATP* model which would expand upon the *CAPM* by adding addition regressors. A fourth example is the equation presented by Fama & Schwert (1977) when seeking an answer to the question of whether investment in stocks hedged against inflation.

The chapter begins by reviewing a paper that models the premia for forward exchange rates. The remaining sections present estimates of the models suggested above. In each case, I have chosen to estimate the variable coefficients as random walks; other models are feasible, but the purpose of this chapter is to illustrate the many applications of the Kalman filter in econometrics rather than to present a complete analysis of the models at hand.

133

7.2. Modeling the forward exchange rate premia

The paper reviewed here attempts to model something that cannot be observed. Wolff (1987) presents a model of the unobservable premia for the forward exchange rate. Conceptually, the forward exchange rate should equal the expected value of the spot rate at the future date plus a risk premium, $P(t)$. As the expectation is uncertain, Wolff defines the difference between the actual spot rate at time $t + 1$, $S(t + 1)$, and the expectation of this spot rate made at time t to be a normally distributed stochastic variable, $v(t + 1)$. The model used is as follows:

$$F(t, t + 1) - S(t + 1) = P(t) + v(t + 1) \qquad (7.1)$$

$F(t, t+1)$ is the forward rate at time t for a transaction at time $t+1$. This equation is similar to my equation (5.1) with $y_t \equiv F(t, t+1) - S(t+1)$, $x_t \equiv P(t)$, $C_t \equiv 1$, $\Gamma \equiv 0$ and $\varepsilon_t \equiv v(t+1)$. The unobserved vector of risk premia $P(t)$ is modeled either as a first order autoregressive process as in equation (5.2) or as a first order moving average process. The hyperparameters — the variances of the error terms in (7.1) and in the state transition equation (5.2) and the unknown parameter in the autoregressive or moving average representation of the latter — are estimated using the maximum likelihood methods outlined in Chapter 5. Wolff's conclusion is that "more than half the variance in the forecast error that results from the use of current forward rates as predictors of future spot rates is accounted for by the variation in premia. ... [T]he methodology can be applied straightforwardly to similar problems in the context of different financial markets." (Wolff (1987), p. 405) The remainder of this Chapter will illustrate the validity of his final assertion.

7.3. Modeling the short term risk free rate of interest

Chan *et al.* show that many models of the risk free rate of interest, r_t, may be specified as follows:

$$r_t - r_{t-1} \equiv \Delta r_t = \alpha + \beta r_{t-1} + \varepsilon_t \qquad (7.2)$$

The model is completed by specifying that the term structure has a zero mean and a variance of ε_t:

$$\text{var}(\varepsilon_t) = \sigma^2 r_t^{2\gamma} \qquad (7.3)$$

With $\gamma > 0$, the model exhibits a time–varying term structure. This model — with $\gamma = \frac{1}{2}$ — has been extensively examined in Cox *et al.* (1985). Chan *et al.* compare different restrictions and find that the most viable models are those with $\gamma \geq 1$. It would take us too far afield to estimate γ in the present context. Rather I will present two different estimations. One will assume that $\gamma = 0$; the second will set γ to unity and divide all variables in equation (7.2) by r_t.

Why should the parameters α and β be constant? Chan *et al.* take care in explaining how different authors have applied zero restrictions to these two terms, but they do not explain why they should be constant. I suggest that these two parameters are in essence variable; at least the hypothesis of stable coefficients should be tested prior to estimation as outlined in Chapters 2 and 3. Here I will estimate the following model:

$$\Delta r_t = \alpha_t + \beta_t r_{t-1} + \varepsilon_t \tag{7.4}$$

$$\alpha_t = \alpha_{t-1} + \xi_{1t} \tag{7.5}$$

$$\beta_t = \beta_{t-1} + \xi_{2t} \tag{7.6}$$

The variables Δr_t and r_t are considered exogenous data for our purposes here. The three variance terms are in one formulation assumed constant; in the second variation, $\text{var}(\varepsilon_t)$ is assumed to be heteroscedastic and proportional to the square of r_t. This specification of the model is estimated by first dividing through by r_t and then estimating as usual. If $\text{var}(\xi_{1t})$ and $\text{var}(\xi_{2t})$ estimate as non-zero, then the implicit hypothesis of constant coefficients in (7.2) should be rejected.

The variable used for the risk free rate is the one month U.S. Treasury Certificates of Deposit rate (CD). The period of estimation is January 1972 to November 1990.

7.4. Estimating the spread

In this section we examine the spread between the three and six month U.S. Treasury bill rate. The model used is based on the *expectations hypothesis* for the term structure. Basically, the hypothesis states that the longer interest rate would be an average of the current short rate and the *expected* short rate for the next period plus a term premium. Mankiw & Miron (1986) present this hypothesis as

$$R_t = \theta + \frac{1}{2}\left(r_t + E\{r_{t+1}|\Omega_t\}\right) \tag{7.7}$$

The upper case R_t represents the longer — the six month — rate in period t while the lower case r_t is the shorter — the three month — rate again at time t. $E\{x_t|\Omega_{t-1}\}$ is expectation for x_t taken with the information available at time Ω_{t-1}. In equation (7.7), θ is a term premium.

The $E\{r_{t+1}|\Omega_t\}$ is not observable. It may, however, be approximated by the actual rate at $t+1$ minus an error term implying that the expectation plus a forecast error equals the actual rate:

$$r_{t+1} = E\{r_{t+1}|\Omega_t\} + \varepsilon_t \qquad (7.8)$$

Mankiw & Miron specify the term premium as constant. This is not justifiable except perhaps in the very short term which the 20 years that encompass their estimation period certainly is not. The other half of the right hand side of equation (7.7) is the average of two interest rates. If the expectations hypothesis is correct, the coefficient of this term should indeed be constant. If it estimates as a variable, then one should question the validity of the basic model.

To express the model in (7.7) and (7.8) in the now familiar state space form, subtract r_t from both sides of (7.7) and solve for the difference $r_{t+1} - r_t$. Then, using (7.8) and allowing the coefficients in (7.7) to vary according to a random walk, we get the following model:

$$r_{t+1} - r_t = \alpha_t + \beta_t (R_t - r_t) + \varepsilon_t \qquad (7.9)$$

$$\alpha_t = \alpha_{t-1} + \xi_{1t} \qquad (7.10)$$

$$\beta_t = \beta_{t-1} + \xi_{2t} \qquad (7.11)$$

The two variables are $(r_{t+1} - r_t)$ and $(R_t - r_t)$. Given the expectations hypothesis, we would expect var(ξ_{2t}) to be zero and that $\beta_t = 2$. The term premium, $-\frac{1}{2}\alpha_t$, could well be variable even under the expectations hypothesis. We therefore expect even var(ξ_{1t}) to be positive. The data used are the 6–month (R_t) and the 3–month (r_t) U.S. Treasury bill rates from January 1959 to November 1990.

7.5. Securities as a hedge against inflation

Fama & Schwert (1977) discuss the relationship between the nominal returns to securities and the rate of inflation. This latter variable may be divided into two parts: the expected rate plus a forecast error. The question to be answered is to what extent security returns hedge against inflation. Letting r_{it} stand for the returns to security i at time t and $\dot{p}_t$ represent the rate of inflation at time t, Fama & Schwert's main model can be written as follows:

$$r_{it} = \alpha_i + \beta_i E\{\dot{p}_t|\Omega_{t-1}\} + \gamma_i (\dot{p}_t - E\{\dot{p}_t|\Omega_{t-1}\}) + \varepsilon_{it} \qquad (7.12)$$

If $\beta_i = 1$ then the nominal returns to security i are a hedge against *expected* inflation. If $\gamma_i = 1$ then the returns are a hedge against *unexpected* inflation.

Why the obsession with *complete* hedges? While the Fisher equation may explain the relationship between nominal returns and real returns *in equilibrium*, most of our observations would seem to be out of equilibrium. By estimating a variable coefficient model, one has the option of tracing the history of the *partial* protection against inflation, an exercise which should be just as interesting as a "once and for all" test of the theory.

Fama & Schwert use the one month treasury bill rate, called R_t in the system below, to proxy the unobservable expected rate of inflation. Once again, the coefficients of equation (7.12) will be assumed to follow a random walk. The model becomes:

$$r_{it} = \alpha_{it} + \beta_{it}R_t + \gamma_{it}(\dot{p}_t - R_t) + \varepsilon_{it} \tag{7.13}$$

$$\alpha_{it} = \alpha_{it-1} + \xi_{1t} \tag{7.14}$$

$$\beta_{it} = \beta_{it-1} + \xi_{2t} \tag{7.15}$$

$$\gamma_{it} = \gamma_{it-1} + \xi_{3t} \tag{7.16}$$

The data are the monthly returns to the Standard and Poor (S&P) Composite Index (r_t), the monthly rate of change in the U.S. CPI $(\dot{p}_t)$, and the 30–day CD rate (R_t) for January 1970 through November 1990.

7.6. A simple "APT" model

The last financial model presented in this section is a simple version of the *Arbitrage Pricing Theory* popularly known as the *APT* model. In contrast to the *CAPM*, the *APT* model seeks other factors, in addition to the single "market risk" — m_t — that the *CAPM* hypothesizes, that effect the risk in holding a given asset. Berry *et al.* (1988) note a number of risk factors. One of these is used here: the risk that the term structure of interest rates may change. This factor , ΔR_t, is here approximated by the difference in the returns between the 10–year government bond and the 30–day CD rate. The model presented by Berry *et al.* relating the returns to a security r_{it} to the two risk factors is

$$r_{it} = \alpha_i + \gamma_i \Delta R_t + \beta_i m_t + \varepsilon_{it} \tag{7.17}$$

Not surprisingly, the assumption of constant coefficients is questionable. The discussion in Chapter 1 on the stability of *beta* is relevant here. By now, consensus should be that this coefficient is time–varying. Along the same lines, the coefficient attached to the risk for a changing term premium is only constant by approximation. Applying the usual random walk to the parameters, we obtain the model to be estimated:

$$r_{it} = \alpha_{it} + \gamma_{it}\Delta R_t + \beta_{it}m_t + \varepsilon_{it} \tag{7.18}$$

$$\alpha_{it} = \alpha_{it-1} + \xi_{1t} \tag{7.19}$$

$$\gamma_{it} = \gamma_{it-1} + \xi_{2t} \tag{7.20}$$

$$\beta_{it} = \beta_{it-1} + \xi_{3t} \tag{7.21}$$

The data used will be the S&P Utility index (r_{it}), the S&P Composite index (m_t), the 10–year U.S. Government bond and the 30–day CD rate for January 1970 to November 1990.

7.7. A macro model: the marginal propensity to save

This section addresses a function that has been much discussed in the literature. Indeed, we are probably climbing out on a rather thin limb in presenting this model. There are a number of problems involved, not the least being the simultaneous relationship of savings and disposable income. However, as the model has useful implications for other application of the technique presented here, we venture into potentially dangerous territory.

The basic hypothesis will be that savings is a function of permanent income. Savings is defined by the difference between private consumption and disposable personal income (dpi); permanent income is proxied by dpi lagged one quarter. This also allows us to dodge the simultaneity problem. Introductory texts in Economics present the marginal propensity to save, mps, as a constant. There is no reason for this assumption. Therefore we will assume it to be variable and represent it by the state variable.

The data used — Swedish quarterly data from 1963 to 1990 — are not seasonally adjusted. As there is no justification for a constant season pattern, we will also estimate seasonality as a variable coefficient. The model estimated will then be equations (4.1) and (4.2) where the state will be five–dimensional:

$$x_t = [\alpha_t \quad \beta_t \quad \delta_{1t} \quad \delta_{2t} \quad \delta_{3t}]'$$

Here the δ_{it}'s are the seasonal factors while α_t is the constant in the savings function and β_t is the mps. Note that the state variances — the ξ's — are also five–dimensional. However, the three variances for the season terms are constrained to be equal.

The seasonals used are trigonometrical (See Harvey (1989), p. 173). Thus we have that $\Phi_{33} = \Phi_{44} = \cos(\frac{\pi}{2})$ and $\Phi_{34} = -\Phi_{43} = \sin(\frac{\pi}{2})$. The Φ matrix is then time invariant and appears as follows:

$$\Phi = \begin{bmatrix} 1 & 0 & 0 & 0 & 0 \\ 0 & 1 & 0 & 0 & 0 \\ 0 & 0 & 0 & 1 & 0 \\ 0 & 0 & -1 & 0 & 0 \\ 0 & 0 & 0 & 0 & -1 \end{bmatrix}$$

C_t, however, will be time–varying. A typical row will be:

$$C_t = [1 \quad x_t \quad 1 \quad 0 \quad 1]$$

TABLE 7.1. **Tests for constant parameters.** Except for the fluctuations test, the entries in the table are the probability that the null hypothesis is true.

Model	T^a	White[b]	Pagan[c]	Fluct[d]	Arch[e]	L–B[f]	G–Q[g]
Longstaff I[h]	220	$1.12E-3$	$6.0E-31$	1.28	$4.90E-4$	$2.3E-12$	$9.90E-3$
Longstaff II[i]	220	$7.41E-2$	$1.29E-4$	1.34	$1.29E-3$	$1.27E-7$	$5.29E-3$
Mankiw[j]	140	$9.51E-2$	$8.85E-2$	0.87	$1.90E-1$	$5.85E-2$	$6.61E-2$
APT[k]	244	$4.31E-1$	$4.72E-2$	1.42	$1.83E-2$	$1.71E-1$	$3.84E-3$
Fama[l]	251	$1.99E-2$	$1.57E-2$	2.42	$1.07E-1$	$7.67E-3$	$6.83E-1$
Savings	108	$4.97E-4$	$1.83E-1$	1.13	$6.4E-15$	0.00^m	$2.71E-3$

[a]The number of observations.
[b]White's test (White, 1980).
[c]The Breusch–Pagan test(Breusch & Pagan, 1979).
[d]The fluctuation test. The 90% level is 1.35 for $T \geq 220$ and 1.42 otherwise.
[e]The *ARCH* test (White, 1980).
[f]The Ljung–Box test.
[g]The Goldfeld–Quandt test.
[h]Equation (7.2) with homoscedastic residual variance: $\gamma = 0$ in equation (7.3).
[i]Equation (7.2) with heteroscedastic residual variance: $\gamma = 1$ in equation (7.3).
[j]Equation (7.9).
[k]Equation (7.18).
[l]Equation (7.13).
[m]The seasonality causes this zero value: $1.3E-104$ is about zero

7.8. Testing for parameter stability

The above models should be subjected to the same battery of tests outlined in Chapter 2. Table 7.1 presents the basic tests for heteroscedasticity. As expected, at least one of the tests for each of the models rejects the null hypothesis of homoscedasticity or of constant coefficients. Even the *Longstaff* model with a heteroscedastic residual variance — where $\gamma = 1$ in equation (7.5) — shows evidence of non constant coefficients as the *Breusch–Pagan*, the *ARCH* and the *Goldfeld–Quandt* tests are positive. I note that the *Savings* model is tested as the usual function with seasonal dummies.

While the statistics are not reported, all of the models show evidence of at least one structural break when the recursive residuals are tested using the *cusumsq* test. This is not surprising given the length of the data in the samples. This does, however, raise the question of whether the non constant model parameters are best described by continuous functions or by piece-wise constant ones. I will return to this issue below.

TABLE 7.2. **Random walk models.** Only one specification of the time–varying parameters has been considered.

	Long–I[a]	Long–II[b]	Mankiw	APT	Fama	Savings
$\mathrm{var}(\xi_{1t})$	$3.27E-2$	$3.21E-2$	$1.22E-1$	$1.33E-5$	$2.58E-3$	$6.64E-16$
$t\text{-}value$	$1.16E+7$	$1.93E+1$	$2.10E+0$	$1.74E+1$	$2.47E+1$	$1.24E-4$
$\mathrm{var}(\xi_{2t})$	$4.85E-3$	$4.84E-3$	$2.06E-1$	$2.17E+0$	$1.06E-1$	$6.02E-5$
$t\text{-}value$	$1.72E+6$	$6.43E+1$	$2.33E+0$	$1.12E+1$	$1.19E+1$	$1.58E+1$
$\mathrm{var}(\xi_{3t})^c$				$4.85E-4$	$1.88E-6$	$3.07E-3$
$t\text{-}value$				$1.42E+1$	$1.02E+2$	$3.09E+1$
$\mathrm{var}(\varepsilon_e)$	$6.72E-7$	$6.99E-5$	$1.16E-1$	$6.28E-4$	$1.15E-3$	$6.38E-2$
AIC^d	$6.97E-7$	$7.24E-6$	$1.18E-1$	$6.60E-4$	$1.20E-3$	$6.87E-2$
LR^e	$1.60E+2$	$5.58E+1$	$1.40E+1$	$1.62E+2$	$1.47E+2$	$2.66E+2$
T^f	220	220	376	244	244	108
$R^2_{t\mid t-1}{}^g$	0.999	0.998	0.661	0.538	0.199	0.969
$R^2_{t\mid t}{}^h$	0.999	0.999	0.893	0.597	0.324	0.988
Q^i	$1.61E-4$	$1.64E-4$	$7.74E-12$	$6.38E-2$	$3.26E-1$	$3.60E-1$
$Hetero^j$	$2.26E-2$	$2.39E-2$	$3.52E-42$	$3.55E-3$	$6.51E-1$	$5.34E-1$
$Arch^k$	$2.43E-2$	$2.45E-3$	$1.02E-10$	$2.41E-3$	$3.52E-1$	$9.14E-1$
$v\,Neum^l$	$1.30E-1$	$1.33E-1$	$5.80E-3$	$3.04E-3$	$1.94E-3$	$6.62E-1$
$recur\text{-}t^m$	$9.73E-1$	$9.80E-1$	$7.34E-1$	$1.81E-1$	$8.00E-1$	$7.97E-1$
$cs^{2\,n}$	$4.25E-6$	$4.43E-6$	$1.34E-25$	$4.75E-6$	$4.20E-6$	$1.16E-2$

[a]The Longstaff model with homoscedastic residuals.
[b]The Longstaff model with heteroscedastic residuals.
[c]For the *Savings* model, this term is the variance of the seasonal components.
[d]Akaike information criterion.
[e]Likelihood ratio.
[f]Number of observations.
[g]R^2 calculated from the recursive residuals.
[h]R^2 calculated from the updated states.
[i]P–value for the Ljung–Box test.
[j]P–value for the Goldfeld–Quandt test. See 2.2.2.2.
[k]P–value for the *ARCH* test.
[l]P–value for a two-sided von Neumann test.
[m]P–value for the recursive *t–test*.
[n]P–value for the *cusumsq* test.

7.9. The estimated models

Table 7.2 presents the estimates of the models outlined above. With the exception of Fama's model, all of the estimates indicate a structural break somewhere in the sample. While I have not pursued the issue here, better estimates may be obtained by splitting the sample into two or more subsam-

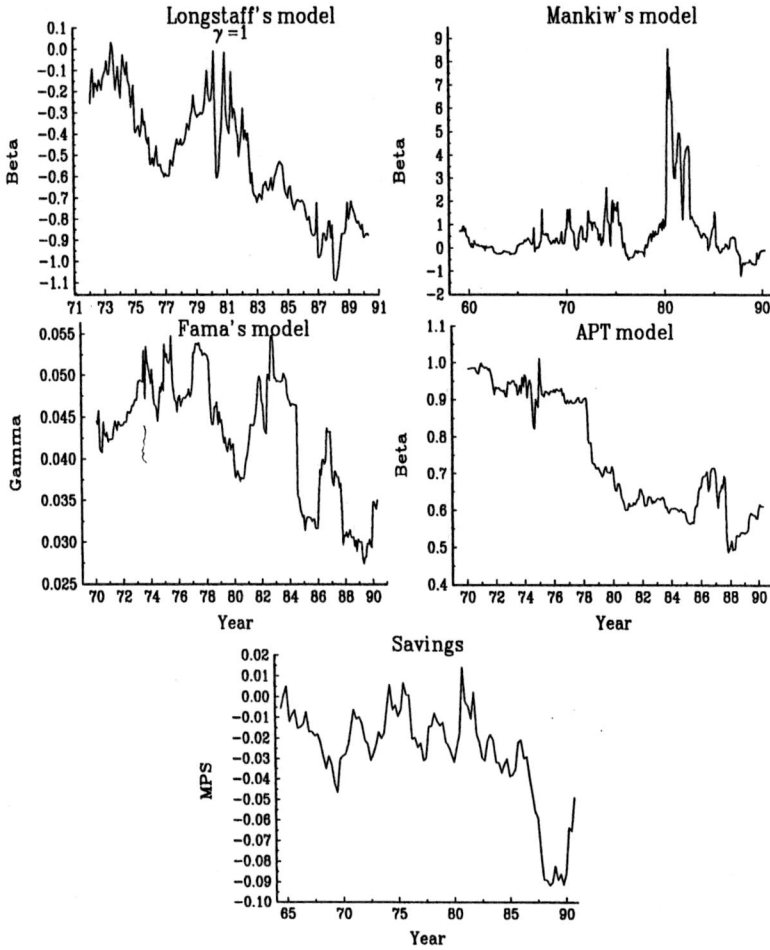

Figure 7.1. The estimated states for the different models.

ples before estimation. Further, other models for the variable coefficients than the *random walk* model could be estimated. For example, a *random coefficients* model, estimated for the different subsamples, could well be interpreted as evidence for piece-wise constant "steady–state" values for the coefficients.

It is of some interest to compare the estimates of the states above to the estimates published by Chan *et al.*, Mankiw & Miron and Fama & Schwert. As the exact data used in these three papers is not available locally, the estimates are not exactly comparable. However, we would expect the estimates to be approximately the same. Further, it is not at all obvious how to compare the state vector of the models presented here with the single coefficient in the published papers. One obvious choice is the *final* value of

TABLE 7.3. **Comparing results.** The final value of the state vector as well as its average is compared to the estimated coefficient presented by the respective authors.

Author	Coefficient	Published estimate	Final State	Average State
Longstaff I[a]	β	-0.18	-0.51	-0.88
Longstaff II[b]	β	-0.31	-0.51	-0.88
Mankiw[c]	β	0.46	-0.10	0.62
Fama[d]	β	-5.52	-8.59	-11.68
Fama	γ	-0.77	0.04	0.42

[a]See Chan *et al.* (1992), p. 1218.
[b]See Chan *et al.* (1992), p. 1218.
[c]See Mankiw & Miron (1986), p. 217.
[d]See Fama & Schwert (1977), p. 128.

the state; however, even the average value of the state may be compared with the single estimate presented in these three papers. The results are summarized in Table 7.3.

By in large, the estimates are similar. The main exception seems to be the γ-coefficient in equation (7.14). Fama & Schwert's results were that common stocks did not seem to hedge against inflation. My estimates — for a later period but using American data[1] — indicate the they were indeed a hedge against inflation: all of the γ-values in equation (7.16) are positive.

7.10. Bringing it all together

This book has considered models with time–varying parameters. An implicit hypothesis advanced has been that techniques for estimating such models should well be in every econometrician's tool box. The main example used is the Capital Asset Pricing Model, the *CAPM*, from the theory of finance. This model provides the perfect example for demonstrating the techniques needed to estimate models with variable parameters: it is well known and widely accepted and there is an abundance of data to experiment on. The data used here is for the Stockholm Exchange and ranges from January 1971 through December 1991.

I began with a brief history of "*beta*", the regression coefficient in the *CAPM*. Markowitz introduced the model in 1959; Fama, almost a decade later, seems to have christened the coefficient. Initially, *beta* was assumed to

---

[1]I use the Standard & Poor Composite Index while Fama & Schwert use the New York Stock Exchange value weighted index.

be constant but in the early 1970's, constancy became *piecewise* constancy. Blume began a discussion on transition matrices that was expanded upon throughout the 1970's. Towards the end of the decade, more attention was directed into finding a model for a time–varying *beta* instead of a constant one with discrete shifts. Fabozzi & Francis assume a random coefficients model for *beta* and design a test to discover if their assumption is justified; Sunder tests whether *beta* follows a random walk; Bos & Newbold estimate a mean reverting model. No matter which model one prefers, the evidence seems to point to models with varying rather than constant coefficients.

The next two chapters present and perform a number of tests for parameter constancy. The models presented have two different types of coefficients. One type is the regression coefficients themselves, that is, *beta* and the intercept in equation (1.5). The other category consists of the *hyperparameters* of the model. These are basically the variances of the in equations (5.1) and (5.2), although there may be others such a "steady–state" value of a coefficient or an element in a transition matrix. The models estimated here have assumed that this latter type of coefficient is constant — or at least piecewise constant — and that the modifier "time–varying" applies to the first category of coefficients.

Tests for heteroscedasticity in the regression model will reject the null hypothesis of constant residual variance if the coefficients of the model are variable. Thus White's, Breusch–Pagan's and Goldfeld–Quandt's tests as well as the *ARCH* and the Ljung–Box tests may be used to test for the adequacy of a model with constant coefficients. The *fluctuations test* is a more direct test for the adequacy of the usual regression model. Sunder's and LaMotte & McWhorter's tests allow one to ask more specific questions as to the nature of the time–varying coefficients.

The *cusumsq* test together with the lesser known *mosumsq* test provides a general test for the stability of the hyperparameters. Shifts in these are interpreted as structural shifts. As the period studied is rather long, one would expect to find some evidence of structural breaks in the data. And so indeed is the case. Only one of the 21 stocks shows no evidence of a shift in the hyperparameters. However, this stock — Pharos — rejects the null hypothesis of constant regression coefficients using the *fluctuations* test.

The final test presented is the *Flexible Least Squares* test. Unlike the others, this *FLS* is not really a test but a method of describing the data. By placing different *a priori* weights on the deviations of the states from period to period and on the residuals in the measurement equation, one can get a feeling for the data. In particular, the consequences of the assumption of constant parameters on the regression equation is highlighted. With little or no restriction on coefficient variation, the system tracks the actual data quite well. If there is a rather heavy penalty on coefficient variation, the

FLS solution approximates the *OLS*. However, if there remains evidence of coefficient variation in spite of a relatively high cost placed on such variations, then one finds evidence in the data that constant coefficient models are inadequate.

I have assumed that the sin of *commission* is rather less than the sin of *omission*. Rejecting a true null hypothesis of constancy has smaller repercussions on further analysis than not rejecting a false one. I have therefore used the 10% level as the critical value in the tests. I have thus chosen to reduce the possibility of not rejecting a false null and willingly run the risk of rejecting a true one. I do many tests: if one says "reject", then I reject the null. This is a rather conservative strategy that follows from my assumption that doing something is better than doing nothing.

There are of course some problems here. The tests are not independent and I have not even attempted to guess the probability that some reject and others do not. Further, if the null is true and I estimate a model assuming that it is false, I will get a poorer model than if I had estimated under the assumption the null was true. One encounters, however, suggestions that the pretesting level should be "much" higher than the usual 5% — perhaps 25% or even 50%. (See Maddala (1989), p. 24) Thus my choice of 10% is not all that conservative.

Chapter 4 presents the Kalman filter which is really a variant of exponential smoothing where the parameters are allowed to be time–varying. The chapter concludes with a brief description of the implementation of the filter in estimating *beta*. In this chapter, I have presented both a formal derivation of the filter as well as an example of its use. The filter is presented for *univariate* systems as equations (4.8) – (4.14). If the matrix C_t is time–invariant, the equations may be solved off-line for a steady–state value. Equation (4.24) illustrates the "filtering" involved: the estimation of the state variable at time t conditional on the information then available is a weighted average of the estimation of the same state conditional on the information available in the previous period and the new information. If a steady–state value is used, then the filter actually is the traditional exponential smoother. However, as the applications in this book are not time–invariant, it is not possible to solve for a steady–state value of the Kalman filter. Equation (4.14) shows why: at each step, the current value of the C_t matrix is required to calculate the filter. We even face an additional problem: the hyperparameters of the systems are not known and must be estimated.

The heart of the book is really the chapter on estimation. While it has been theoretically possible to estimate models with time–varying coefficients since the early 1970's, the cost of computer and the programming effort required have been effective barriers until the last few years. Chapter

5 presents the models to be estimated, the tools needed to estimate them, as well as the actual estimations for the 21 stocks included in the sample.

The likelihood function, equation (5.14), is maximized using the numerical routines provided by GAUSS. The variance-covariance matrix of the estimated parameters is the heteroscedastic consistent one suggested by White (1982) and implemented in the GAUSS routines.[2] The estimates were also subjected to a battery of diagnostic tests. By in large, in those cases where the tests were unsatisfactory, the main fault lay in structural instability. Once the data was divided into different subsamples, the estimates improved dramatically. However, the tables in Chapter 5 present only estimates that span the entire sample.

The tests in Chapter 2 indicated that about 70% of the models should estimate as *RW*s and about 5% as *RCF*s. (See page 44.) The estimates, however, do not follow this proportion. About 50% — 10 of 21 if the *MM* model is included amongst the random walks — estimate as *RW*s and about 24% — 5 of 21 — as *RCF*s. Also fewer than expected estimate as random walks when the sample was truncated in an attempt to find structurally stable parameters. However, my sample of 21 stocks is not only small, but also far from being random. Thus too much weight should not be placed on these percentages.[3]

Chapter 6 illustrates how the estimates may be improved by dividing the sample into smaller periods which show evidence of structural stability. Using the *cusumsq*, and to a lesser extent, even the *mosumsq* statistic, periods with no evidence of changes in the hyperparameters may be found. The partitioning of the sample is more an art than a science. At times, an extremely large residual will seem to indicate a shift when it is in fact caused by a more or less random disturbance. On the other hand, a large residual may well indicate that a new epoch is here. One must sort each case on its own merits.

A second point raised in Chapter 6 is that different specifications of the model for *beta* yield different time paths for that coefficient and that the correlation between them may rather low. In particular, the time path for *beta* in the models that specify a constant "steady–state" value of the parameter is poorly correlated with that from models that allow the mean to vary. Thus one must exhibit care in interpreting the model: the "historical" development of *beta* is very much a function the model describing it.

[2]If the reader does not own the MAXLIK routine, there are others available for the effort required in downloading them. These are on a gopher server maintained by the Department of Economics at American University. The following *URL* will attach you to the archive:

gopher://gopher.american.edu/11/academic.depts/cas/econ/software/gauss

[3]The data from Table 5.1 are more in line with the findings from the pretests. There 45% are *RW*s and 19% *RCF*s.

The final chapter has presented a number of possible applications of the methods presented here. The models in this chapter are by no means definitive; they are intended to serve as examples of applications of time–varying parameter models. Especially the model of the savings function should be reformulated to take into consideration the simultaneous nature of the data.

However, the main point made in not only the final chapter but also in the book is that many time–series models estimate very satisfactorily with time–varying parameters. While maximum likelihood estimation of models specified using the Kalman filter are no where near as simple to estimate as *OLS* models, today's state of the art in computers and programming have reduced the effort and cost required considerably; thus other considerations than computational difficulty should dictate which type of model one should estimate. This study has presented six different models, five financial and one macro. The evidence indicates that all of these could well be estimated using variable coefficient methods. To say that *all* models should have such coefficients is to make the same mistaken generalization as those who advocate constant coefficient models as the main rule. Still, my conclusion must be that many more models than those few which appear in today's journals should be estimated with time–varying coefficients.

Appendix A

Tables of References

There are quite a few references to earlier work on estimating *beta* in the main text. The following four tables try to present a short summary of these references.

A.1. Stability tests by partitioning data

Table A.1 lists a few of the articles that considered a *piece-wise* variable *beta*. The method used was to partition the data into different intervals and check on stability between the periods. The second column states the length of the partition in months; the final column summarizes the conclusions drawn.

TABLE A.1. **Testing stability using exogenous partitions.**

Reference	Estimated period	Conclusion
Blume (1971)	84	stable
Gonedes (1973)	84	stable
Altman *et al.* (1974)	1 – 52	stable for longer periods
Baesel (1974)	12 – 108	unstable
Klemkosky & Martin (1975)	60	unstable
Roenfeld *et al.* (1978)	48	stable for longer periods
Gheysens *et al.* (1979)	24	stable; data is weekly
Alexander & Chervany (1980)	1, 2, 4, 6, 9	stable for extreme pentads
Theobald (1981)	up to 220	stable for longer periods
Hansson & Wells (1987)	60	unstable

TABLE A.2. **Testing stability using endogenous partitions.**

Reference	Criterion	Conclusion
Kon & Jen (1978)	bull/bear	unstable
Fabozzi & Francis (1979)	bull/bear	unstable
Kon & Jen (1979)	discrete shifts	unstable
Miller & Gessis (1980)	discrete shifts	unstable

The second table, Table A.2 presents the articles that have used some endogenous criterion such as bull/bear market for partitioning the data. The second column states the type of partition; the final column summarizes the conclusions drawn.

A.2. Tests for heteroscedasticity

That there might be heteroscedasticity in the market model was an issue that occupied researchers especially during the later half of the 1970's. Table A.3 lists the references in the book that have tested for *beta* stability by testing for heteroscedasticity in the *CAPM*. The second column indicates which stock exchange is considered; the final column summarizes the conclusions drawn by indicating the percentage of the tests that reject the null hypothesis of homoscedasticity.

TABLE A.3. **Tests for heteroscedasticity.**

Reference	Exchange	Rejection
Martin & Klemkosky (1975)	NYSE	8.4
Morgon (1976)	NYSE	58.8
Brenner & Smidt (1977)	NYSE	17.5
Brown (1977)	NYSE	35.0
Bey & Pinches (1980)	NYSE	21 - 33
Collins et al. (1987)	NYSE	30 - 60
Knif (1989)	Finnish	49 - 85
Wells (1990)	Swedish	28 - 42

TABLE A.4. **Models in the literature.** This table lists the references in the book to the different models. The final column indicates whether the cited article tests for the model indicated or in fact estimates the model.

Model	Reference	Estimate or test
Mean Reverting	Schaefer *et al.* (1975)	test
	Ohlson & Rosenberg (1982)	estimate
	Bos & Newbold (1984)	test
	Collins *et al.* (1987)	estimate
	Knif (1989)	estimate
	Faff *et al.* (1992)	test
	Wells (1994)	estimate
Random Coefficients	Schaefer *et al.* (1975)	test
	Fabozzi & Francis (1978)	test
	Fabozzi *et al.* (1982)	test
	Simonds *et al.* (1986)	test
	Knif (1989)	test
	Brooks *et.al.* (1992)	test
	Wells (1994)	test
Random Walk	Fisher (1971)	estimate
	Kantor (1971)	estimate
	Szeto (1973)	estimate
	Schaefer *et al.* (1975)	test
	Sunder (1980)	test
	Garbade & Rentzler (1981)	estimate
	Alexander *et al.* (1982)	test
	Fisher & Kamin (1985)	estimate
	Simonds *et al.* (1986)	estimate
	Knif (1989)	estimate
	Wells (1994)	estimate

A.3. Models in the literature

Table A.4 lists the references to the different models. I have included the *ROS* and the *MM* models under the *mean reverting* listing to keep the table manageable.

Appendix B

The programs and the data

The programs used in this study are available for downloading via anony-
mous *ftp* at our server ftp.ec.lu.se on the subdirectory /pub/nek/kalprog.
These include all routines except the basic maximum likelihood routine
which must be purchased from Aptech Systems. Those routines with the *g*
extension are suitable to be placed in the *src* subdirectory under the main
GAUSS directory; however, some use globals defined in other programs.[1]
The routines with the *prg* extension are programs that must be executed.
The *fls*, *lamotte* and the *kalr* programs are main routines. This section
describes how to use the programs. The data used in the book are also to
be found on the same subdirectory. The Swedish stocks are in a GAUSS
data set *findata.dat* with the labels in *findata.dtx*. The remaining data sets
are in ASCII with the extension *txt*. Files stored with the extension em fmt
are those for the *cusumsq* tests as explained below.

First of all, the data must be entered into the program and compacted,
as none of the routines handle missing values. My data had missing values
at the beginning or the end of the sample but not in the middle of it. Thus
my solution was to search through the data until a non-missing value was
found and that was accepted as the first observation of the series for the
dependent variable; my independent variable spanned the entire series so
there was no problem here. The variables were then concatenated horizon-
tally with the dependent variable as the *last* column of the matrix and then
"packr"'ed. Finally a column of ones was attached as the first column of
the data matrix. The starting value was then used in plotting so that the
horizontal axis could have acceptable labels.

There is one disclaimer that I must issue: while I enjoy programming, I
am by no means a "computer freak" and these programs have been written
for my own use. While the subroutines are easy to read, the main programs

---

[1]Sorry about that. The programs indicate when globals are required.

contain data input routines that have been designed for my particular data set. Note that the ASCII data sets should be entered directly.

B.1. Subroutines

The programs **pagan.g**, **plob.g** and **white.g** take the data matrix as the argument. They produce test statistics and, when appropriate, the number of degrees of freedom for the test. They use no globals.

Recursive residuals are calculated by **recresid.g**. Again, the program takes the data matrix as input. It counts the number of columns in the matrix and uses the last one as the dependent variable. Its output is the vector of recursive residuals.

arch.g does an *ARCH* test and requires residuals — recursive or otherwise — from a regression equation and the number of lags for the autocorrelation function as inputs. It has one output: the test statistic which follows a *chi-squared* distribution with the number of lags in the autocorrelation function as the number of degrees of freedom.

cs2prob.g calculates the probability for the maximal deviation of the *cusumsq* line from the reference line from 0 to 1. This routine takes the maximal deviation and the number of degrees of freedom as the input and produces a *one-sided* probability as the output. For a two-sided test, double the probability that is returned.

cumsumm2.g calculates the *cusumsq* line. It requires as input the recursive residuals, the number of parameters and the constant used in calculating the width of the confidence interval as was explained in the appendix to Chapter 2. It produces three vectors as outputs: the *cusumsq* line, the upper and the lower confidence bounds in that order. Note that the files b*x*.fmt contain the constants *a* from Table 2.4 that may be used with equation (2A.17) to calculate the required constant. The *x* indicates the desired confidence level: 90, 95, 97.5, 99 or 99.5.

mosumm2.g calculates the *mosum*^{2} line. The inputs are the vector of recursive residuals, the number of parameters in the model, the width of the window — that is, the number of observations in it — and the appropriate critical value for the *F*–statistic. As GAUSS does not give the inverse of this statistic, I have used the program *nlsys* from the *Nonlinear equations* module to solve for the critical value.

von_neum.g calculates a number of tests using recursive residuals and the number of estimated parameters as inputs. The tests calculated and printed are the *ARCH*, the Goldfeld–Quandt, the Jacque-Bera, the Kolmogorov-Smirnov, the Ljung-Box and the recursive *t*-test. There are no parameters returned from this subroutine. **vvstat.g** returns similar statistics; use it with the residuals from the maximum likelihood estimates

TABLE B.1. **Program descriptions: the subroutines.** The routines listed in this table are self-contained; they do not require global variables. x is the data input matrix which is T by $k + 1$ with the dependent variable in the last column and the k independent variables in the first k columns. The constant must be in the first column. Note that the vector v of recursive residuals produced by *recresid.g* is used as a calling argument in some of the other routines.

Program	The call	Comments
arch.g	a = arch(v,nlags)	a is $\chi^2(nlags)$
cs2prob.g	pr = cs2prob(mc,df)	pr is one-sided.[a].
cumsumm2.g	{c,u,d} = cumsumm2(v,k,c0)	c is the cs^2 line.[b]
mosumm2.g	{m,u} = mosumm2(v,k,w,m0)	m is the $mosum^2$ line.[c]
pagan.g	{p,dp} = pagan(x)	p is $\chi^2(dw)$
plob.g	f = plob(x)	is the test statistic.[d]
recresid.g	v = recresid(x)	v are the recursive residuals.
von_neum.g	s = von_neum(v,k)	s is a dummy argument.[e]
vvstat.g	s = vvstat(r)	s is a dummy argument.[f]
white.g	{w,dw} = white(x)	w is $\chi^2(dw)$

[a]mc is the maximal deviation of the cs^2 line from the line between 0 and 1; df is the number of degrees of freedom. For a two-sided test, double pr

[b]The upper and lower confidence lines are u and d. $c0$ is the constant that determines the width of this confidence interval. See the appendix to Chapter 2. The routine *cumsumsq.g* is similar but requires that the recursive residuals be a global called vv and requires only k and $c0$ as calling arguments.

[c]w is the number of observations encompassing the window. The upper confidence lines is u. m follows a F-distribution with w and T-w-k degrees of freedom. $m0$ is the appropriate critical value of the statistic. The routine *mosumsq.g* is similar but requires that the recursive residuals be a global called vv and requires only k, w and $c0$ as calling arguments.

[d]See Ploberger *et al.* (1989) for confidence levels.

[e]This routine calculates and prints out a number of tests. These are the *ARCH*, the Goldfeld–Quandt, the Jacque-Bera, the Kolmogorov-Smirnov, the Ljung–Box and the recursive t–test.

[f]r is the vector of the residuals from the maximum likelihood estimates.

vvstat.g produces a number of goodness of the fit statistics for the maximum likelihood estimates. It is designed to be used with programs like *mL_rw.prg* defined below. All arguments but one are passed as globals. These are listed at the start of the routine. The one exception is the vector of residuals which is passed as an argument in the call to the routine.

B.2. The main programs

There are three programs that may be downloaded. Two of these do pretesting and one does the maximum likelihood estimation. While the sub-

TABLE B.2. **Main Programs.** The three main routines listed in this table may be downloaded from our *ftp*. Input must be supplied by the user.

Program	Purpose
fls.prg	Estimates the flexible least squares regressions.
lamotte.prg	Performs Sunder's and La-Motte-McWhorter's tests.
ml_rw.prg	Estimates the hyperparameters in the *Random Walk* specification of the *CAPM*.

routines above are general in nature — that is, they may be applied to many different equations — the programs listed here have been tailor made for the problem at hand. In other words, they are designed with reference to the simple regression model with one independent variable and a constant.

All three require that the data be entered in a very special way: as a matrix with the dependent variable in the last column. The program does not check for missing values, so it you have them, be careful.

fls.prg estimates and plots the *flexible least squares* regressions. The program is a GAUSS translation of a FORTRAN program provided by Prof. Tesfatsion (see Tesfatsion & Veitch (1990).) It is extensively commented. The program as stands plots four diagrams at once on the screen, so it helps it you have either a large screen or a video card with a 1024×768 resolution that GAUSS supports. Having both is very helpful.

lamotte.prg performs the LaMotte–McWhorter and the Sunder tests on the data. As it stands, it loops through an entire data set. Like *fls.prg*, this program has many comments so that one may easily follow its workings. The program calculates eigenvalues for a $T \times T$ matrix and this takes time for large matrices on machines with only 8 megs internal memory. However, on a 66 MHz with 16 megs, with $T = 252$, execution times were acceptable.

ml_rw.prg estimates the hyperparameters in the *random walk* specification of an *OLS* with variable coefficients. It performs a number of goodness of fit tests as well as a number of plots. There is no real problem in expanding this program to estimate the other models.

One final disclaimer is in order. While all the programs have been extensively tested, errors have a way of popping up when least expected. Thus there can be no 100% guarantee that there are no bugs in the programs. Use them and enjoy them. I have.

TABLE B.3. **Listing of the Swedish financial data.** "Period" indicates the
period for which the data is available.

Stock	Period	Stock	Period
Aga	1971:1–1991:12	Investor	1971:2–1991:12
Alfa Laval	1971:1–1991:7	JM Bygg	1982:1–1991:12
Anderson	1983:7–1991:12	Korsnäs–Marma	1971:2–1991:12
Argentus	1981:9–1990:4	Marabou	1971:2–1990:10
Aritmos	1971:2–1991:12	Modo	1971:2–1991:12
Asea	1971:1–1991:12	Munksjö	1971:2–1990:11
Astra	1971:1–1991:12	Nobel	1971:2–1991:12
Atlas Copco	1971:1–1991:12	Perstorp	1971:2–1991:12
Bergman & Beving	1976:12–1991:12	Pharmacia	1971:2–1990:7
BGB	1983:5–1991:12	Pharos	1981:2–1991:12
Custos	1971:2–1991:12	Promotion	1971:2–1991:12
Eken	1976:6–1991:12	Pronator	1980:8–1991:12
Electrolux	1971:1–1991:12	Protorp	1981:2–1991:12
Ericsson	1971:1–1991:12	Providentia	1971:1–1991:12
Esab	1971:1–1991:12	Ratos	1975:2–1991:12
Esselte	1971:1–1991:12	Saab	1971:1–1991:7
Euroc	1971:1–1991:12	Sandvik	1971:1–1991:12
Export	1971:4–1991:12	SCA	1971:1–1991:12
Fabege	1979:4–1991:12	Siab	1982:7–1991:12
Gambro	1983:5–1991:12	Skandia	1980:5–1991:12
Garphyttan	1971:1–1991:12	Skanska	1971:1–1991:12
General index	1971:1–1991:12	SKF	1971:1–1991:12
Gullspån	1981:10–1991:12	Stora Kopparberg	1971:2–1991:12
Hevea	1971:2–1991:12	Sydkraft	1971:2–1991:12
Hexagon	1971:1–1990:9	Skåne Gripen	1980:2–1991:12
Hufvudstad	1971:2–1991:12	Trelleborg	1971:2–1991:12
Incentiv	1971:2–1990:6	Volvo	1971:1–1991:12
Industrivärden	1971:2–1991:12	Åkermans	1971:2–1990:12
Infina	1982:8–1991:7	Öresund	1971:2–1991:12

B.3. The data

Tables B.3 and B.4 list the data available for downloading. The GAUSS
database *findata* are the returns to the stocks and the market index —
as defined on page 92 — used in this study. In all there are 57 stocks of

TABLE B.4. **Listing of the data used in Chapter 7.** "Period" indicates the period for which the data is available. Note that all data is monthly except the Swedish which is *quarterly*.

Variable	File	Period
Standard & Poor Composite index	spci.txt	1970:1–1990:11
Standard & Poor Utility index	spui.txt	1970:1–1990:11
Swedish disposable personal income	sdpi.txt	1963:1–1990:4
Swedish total private consumption	stpc.txt	1963:1–1990:4
U.S. Consumer Price Index (CPI)	cpi.txt	1959:1–1990:11
U.S. Government Bond, 10–year rate	bg10.txt	1970:1–1990:11
U.S. Treasury Certificates of Deposit, 1–month rate (CD)	cd1.txt	1970:1–1990:11
U.S. Treasury Bills, 6–month rate	tb6.txt	1959:1–1990:11
U.S. Treasury Bills, 3–month rate	tb3.txt	1959:1–1990:11

varying lengths as indicated in the table. I have used 21 of them in this study but have during my work on this book estimated models for all 56. As the reader who tries his hand at estimating the models discussed here will soon discover, some are simply hopeless to estimate unless one splits the sample on the basis of, say, the *cusumsq* plot. Hexagon is in this category: it goes crazy near the end of the sample, beginning in September of 1989. This stock was removed from the exchange a year later: it appears that the market model is rather inadequate to explain the changes in the returns to stocks that are about to removed from the exchange, due to either a take over by another company or bankruptcy. Ratos that has an observation in excess of five standard deviations from its mean. The data in Table B.4 is provided in 9 ASCII files with the names indicated in the table.

The Swedish financial data has, as mentioned in the Preface, been provided by Findata in Stockholm. The source of the data should be explicitly stated when using the data either   in research or in the classroom. The index variable has been taken from the tables published by Hansson & Frennberg (1992). In some ways, this is a rather unique data set. The material in it has not been generally available and it is provided here for those who wish to duplicate my results or pursue other studies of the Stockholm exchange.

Bibliography

Ackley, G. (1961) *Macroeconomic Theory*, MacMillan, New York.

Alexander, G. J. and Benson, P. G. (1982) "More on beta as a random coefficient", *Journal of Financial and Quantitative Analysis*, **Vol. 17 no. 1**, pp. 27 - 36.

Alexander, G. J., Benson, P. G. and Eger, C. E.,(1982), "Timing decisions and the behavior of mutual fund systematic risk", *Journal of Financial and Quantitative Analysis*, **Vol. 17 no. 4**, pp. 579 - 602.

Alexander, G. J. and Chervany, N. L. (1980) "On the estimation and stability of beta", *Journal of Financial and Quantitative Analysis*, **Vol. 15 no. 1**, pp.123 - 137.

Altman, E. I., Bertrand, J. and Levasseur, M. (1980) "Comparative analysis of risk measures: France and the United States", *Journal of Finance*, **Vol. 27 no. 4**, pp. 1495-1511.

Anderson, B. D. O. and Moore, J. B., (1979) *Optimal Filtering*, Prentice-Hall, New Jersey.

Athens, M. (1972) "The discrete time linear-quadratic Gaussian stochastic control problem", *Annals of Social and Economic Measurement*, **Vol. 1 no. 4**, pp. 449-491.

Athens, M., (1974) "The importance for Kalman filtering methods for economic systems", *Annals of Social and Economic Measurement*, **Vol. 3 no. 1**, pp. 49-64.

Baesel, J. B. (1974) "On the assessment of risk; some further considerations", *Journal of Finance*, **Vol. 27 no. 4**, pp. 1491-1494.

Bauer, R. J. Jr., Hays, P. A. and Upton, D. E. (1980) "Parameter instability in mutual fund portfolios: a shifting regimes test", *Quarterly Journal of Business and Economics*, **Vol. 26 no. 1**, pp. 50-62.

Berry, M. A., Burmeister, E. and McElroy, M. B. (1988) "Sorting out risks using known APT factors", *Financial Analysts Journal*, **Vol. 44 no. 2**, pp. 29-42.

Bey, R. P. and Pinches, G. E. (1980) "Additional evidence of heteroscedasticity in the market model", *Journal of Financial and Quantitative Analysis*, **Vol. 15 no. 2** , pp. 299-322.

Blume, M. E. (1971) "On the assessment of risk", *Journal of Finance*, **Vol. 24 no. 1**, pp. 1-10.

Blume, M. E. (1975) "Betas and their regression tendencies" *Journal of Finance*, **Vol. 30 no. 3**, pp. 785-795.

Bollerslev, T., Engle, R. F., and Wooldridge, J. M., "Capital assets pricing model with time–varying covariances", *Journal of Political Economy*, **Vol. 96 no. 1**, pp. 116-131.

Bos, T. and Newbold, P. (1984) "An empirical investigation of the possibility of stochastic systematic risk in the market model", *Journal of Business*, **Vol. 57 no. 1**, pp. 35-41.

Brenner, M. and Smidt, S. (1977) "A simple model of non-stationarity of systematic risk", *Journal of Finance*, **Vol. 32 no. 4**, pp. 1081-1092.

Brooks, R. D., Faff, R. W. and Lee, J. H. (1992) "The form of time variation of systematic risk: some Australian evidence", *Applied Financial Economics*, **Vol. 2 no. ?**, pp. 191-198.

Brown, S. J. (1977) "Heteroscedasticity in the market model: a comment", *Journal of Business*, **Vol. 50 no. 1**, pp. 80-83.

Brown, R. L., Durbin, J. and Evans, J. M. (1975) "Techniques for testing the constancy of regression relationships over time", *Journal of the Royal Statistical Society, series B*, **Vol. 37 no. 2**, pp. 149-192.

Breusch, T. S. and Pagan, A. R. (1979) "A simple test for heteroscedasticity and random coefficient variation", *Econometrica*, **Vol. 47 no. 5**, pp. 1287-1294.

Bryson, A. E., Jr., and Ho, Y-C. (1979) *Applied Optimal Control*, John Wiley and Sons.

Chan, K. C., Karolyi, G. A., Longstaff, F. A. and Sanders, A. B. (1992), "An empirical comparison of alternative models of the short-term interest rate", *Journal of Finance*, **Vol. 47 no. 3**, pp. 1209-1227.

Cox, J. C.,Ingersoll, J. E. and Ross, S. A. (1985) "A theory of the term structure of interest rates", *Econometrics*, **Vol. 53 no. 2**, pp. 385-408.

Collins, D. W., Ledolter, J. and Rayburn, J. (1987) "Some further evidence on the stochastic properties of systematic risk", *Journal of Business*, **Vol. 60 no. 3**, pp. 425-449.

Davies, R. B. (1977) "Hypothesis testing when a nuisance parameter is present only under the alternative", *Biometrica*, **Vol. 64 no. 2**, pp.247-254.

DeJong, D. and Collins, D. (1985) "Explanations for the instability of equity beta: Risk-free rate changes and leverage effects", *Journal of Financial and Quantitative Analysis*, **Vol. 20 no. 1**, pp.73-94.

Durbin, J. (1969) "Tests for serial correlation in regression analysis based on the periodogram of least squares residuals", *Biometrica*, **Vol. 56 no. 1**, pp. 1-15.

Edgerton, D. and Wells, C. (1994) "Critical values for the *cusumsq* statistic in medium and large sized samples", *Oxford Bulletin of Economics and Statistics*, **Vol. 56 no. 3**, pp.355-365.

Engle, R. F. (1982) "Autoregressive conditional heteroscedasticity with estimates of the variance of United Kingdom inflation", *Econometrica*, **Vol. 50 no.4**, pp. 987-1007.

Engle, R. F., Lilien, D. and Robins, R. (1987) "Estimating time varying risk premia in the term structure: the ARCH-M model", *Econometrica*, **Vol. 55 no. 2**, pp. 391-408.

Fabozzi, F. J. and Francis, J. C. (1978) "Beta as a random coefficient", *Journal of Financial and Quantitative Analysis*, **Vol. 13 no. 1**, pp. 101-116.

Fabozzi, F. J. and Francis, J. C. (1979) "Mutual fund systematic risk for bull and bear markets: an empirical examination", *Journal of Finance*, **Vol. 34 no. 5**, pp. 1234-1250.

Fabozzi, F. J., Francis, J. C. II, and Lee, C. F. (1982) "Specification error, random coefficients and the risk- return relationship", *Quarterly Review of Economics and Business*, **Vol. 22 no. 1**, pp. 21-31.

Faff, R. W., Lee, J. H. H. and Fry, T. R. L. (1992) "Time stationarity of systematic risk: some Australian evidence", *Journal of Business, Finance and Accounting*, **Vol. 19 no. 2**, pp. 253-270.

Fama, E. F. (1965) "Behavior of stock market prices", *Journal of Business*, **Vol. 38 no. 1**, pp. 34-105.

Fama, E. F. (1968) "Risk, return and equilibrium: some clarifying comments", *Journal of Finance*, **Vol. 23 no. 1**, pp. 29-40.

Fama, E. F., Fisher, L., Jensen, M. and Roll, R. (1969) "The adjustment of stock prices to new information", *International Economic Review*, **Vol. 10 no. 1**, pp. 1-21.

Fama, E. F. and Schwert, G. W. (1979) "Asset returns and inflation", *Journal of Financial Economics*, **Vol. 5 no. 2**, pp. 115-146.

Fisher, L. (1971) "On the estimation of systematic risk", *Proceedings of the Wells Fargo Symposium, July 26-28*, 1971.

Fisher, L. and Kamin, J. (1985) "Forecasting systematic risk: estimates of 'raw' beta that take into account the tendency of beta to change and the heteroscedasticity of residual returns", *Journal of Financial and Quantitative Analysis*, **Vol. 20 no. 2**, pp. 127-149.

Francis, J. C. (1975) "Intertemporal differences in systematic stock price movements", *Journal of Financial and Quantitative Analysis*, **Vol. 9 no. 2**, pp.

Friedman, M. (1957)*A Theory of the Consumption Function, National Bureau of Economic Research, Paper No. 63, general series*, Princeton, N.J.

Garbade, K. (1977) "Two methods for examining the stability of regression coefficients", *Journal of the American Statistical Association*, **Vol. 72 no. 1**, pp. 54-63.

Garbade, K. and Rentzler, J. (1981) "Testing the hypothesis of beta stationarity", *International Economic Review*, **Vol. 22 no. 3**, pp. 577-587.

Gheysens, L., Regidor, B. and Vanthienen, L. (1979) "Cost of equity capital of 185 Belgian companies", *Tijdschrift voor Economie en Manegement*, **Vol. 24 no. 1**, pp. 55-108.

Ghosh, S. K. (1991) *Econometrics: Theory and Applications*, Prentice-Hall International, 1991.

Gonedes, N. J. (1973) "Evidence on the information content of accounting numbers: accounting-based and market based estimates of systematic risk", *Journal of Financial and Quantitative Analysis*, **Vol. 8 no. 2**, pp. 407-433.

Gouriéroux, C., Holly, A. and Monfort, A. (1982) "Likelihood ratio test, Wald test and Kuhn-Tucker test in linear models with inequality constraints on the regression parameters", *Econometrica*, **Vol. 50 no. 1** , pp. 63-80.

Hackl, P. (1980) *Testing the Constancy of Regression Models over Time*, Vandenhoeck & Ruprecht, Göttingen.

Hall, S., Miles, D. K. and Taylor, M. P. (1989) "Modeling asset prices with time–varying betas", *The Manchester School of Economics and Social Studies*, **Vol. 57 no. 4**, pp. 340-356.

Hansson, B. and Frennberg, P. (1992) "Computation of a monthly index for Swedish stock returns: 1919-1990", *Scandinavian Economic History Review*, **Vol. 40 no. 1**, pp. 3-27.

Hansson, B. and Wells, C. (1987) "Investment strategies based on beta coefficients", Beta, **Vol. 1 no. 3-4**, pp. 21-26 (*in Swedish*).

Harvey, A. C. (1989) *Forecasting Structural Time Series Models and the Kalman Filter*, Cambridge University Press, Cambridge, 1989.

Harvey, A. C. (1990) *The Econometric Analysis of Time Series, second edition*, Philip Allen, New York.

Harvey, A. C. (1993) *Time Series Models, second edition*, Harvester-Wheatsheaf, New York.

Harvey, A. C. and Phillips, G. D. A. (1982) "The estimation of regression models with time–varying parameters", in Deistler, M., Furst, E., and Schwodiaur, G., eds., *Games, Economic dynamics and Time Series Analysis*, Physica-Verlag, Würzburg, pp. 306-321.

Hays, P. A. and Upton, D. E. (1986) "A shifting regimes approach to the stationarity of the market model parameters of individual securities", *Journal of Financial and Quantitative Analysis*, **Vol. 21 no. 3**, pp. 307-321.

Jazwinski, A. H. (1970) *Stochastic Processes and Filtering Theory*, Academic Press, New York.

Jensen, M. (1969) "Risk, the pricing of capital assets and the evaluation of investment portfolios", *Journal of Business*, **Vol. 39 no. 1**, pp. 167-247.

Johnston, J. (1984) *Econometric Methods, third edition*, McGraw Hill International, Singapore.

Judge, G. G., Griffiths, W. E., Hill, R. C., Lütkepohl, H. and Lee, T-C. (1985) *The Theory and Practice of Econometrics, second edition*, John Wiley and Sons, London.

Kalaba, R. and Tesfatsion, T. (1989) "An organizing principle for dynamic estimation", MRG working paper # M8818, Department of Economics, University of Southern California.

Kalman, R. E. (1960) "A new approach to linear filtering and prediction problems", *Journal of Basic Engineering, ASME Transactions, Series D*, **Vol. 82**, pp. 35-45.

Kalman, R. E. (1963) "New methods in Wiener filtering", in Bogdanoff, J. L. and Kozin, F., eds., *Proceedings of the First Symposium on Engineering Applications of Random Function Theory and Probability*, John Wiley and Sons, New York, pp.270-388.

Kantor, M. (1971) "Market Sensitivities", *Financial Analysts Journal*, **Vol. 27 no.1**, pp. 64- 68.

Klemkosky, R. C. and Martin, J. D. (1975) "The adjustment of beta forecasts", *Journal of Finance*, **Vol. 30 no. 3**, pp. 1123-1128.

Knif, J. (1988) "Finnish beta coefficients; empirical evidence of instability", *Liiketaloudellinen Aikakauskirja (The Finnish Journal of Business Economics)*,special edition **1988**, pp. 1-17.

Knif, J. (1989) *Parameter variability in the single factor model*, unpublished Ph.D. Thesis, Åbo Akademi.

Kon, S. J. and Jen, F. C. (1978) "Estimation of time- varying systematic risk and performance for mutual fund portfolios: an application of switching regression", *Journal of Finance*, **Vol. 33 no. 2**, pp. 457-475.

Kon, S. J. and Jen, F. C. (1979) "The investment performance of mutual funds: an empirical investigation of timing, selectivity and market efficiency", *Journal of Business*, **Vol. 52 no. 2**, pp. 263-289.

Kon, S. and Lau, W. (1979) "Specification tests for portfolio regressions, parameter stability and the implications for empirical research", *Journal of Finance*, **Vol. 34 no. 2**, pp. 451-464.

Kool, C. J. M. (1989) *Recursive Bayesian Forecasting in Economics: the Multi State Kalman Filter Method*, Drukkerij SSN, Nijmegen

Kool, C. J. M. and Bomhoff, E. J. (1983) "Forecasts with multi-state Kalman filters", appendix 1 in Bomhoff, E. J.,it ed. *Monetary Uncertainty*, North-Holland, Amsterdam.

LaMotte, L. R. and McWhorter, A. Jr. (1978) "An exact test for the presence of random walk coefficients in a linear regression", *Journal of the American Statistical Association*, **Vol. 73 no. 364**, pp. 816-820.

Lee, C. F. and Chen, C. R. (1982) "Beta stability and tendency", Journal of Economics and Business, **Vol. 34**, pp. 201-206.

Litner, J. (1965) "Security prices, risk and maximal gains from diversification", *Journal of Finance*, **Vol. 20 no. 5**, pp. 587-615.

Maddala, G. S. (1989) *Introduction to Econometrics*, MacMillan Publishing Company, New York.

Mandelbrot, B. (1963) "The valuation of certain speculative prices", *Journal of Business*, **Vol. 36 no. 3**, pp. 394-419.

Mandelbrot, B. (1966) "Forecasts of future prices, unbiased markets and martingale models", *Journal of Business*, **Vol. 39**, *special supplement*, pp. 242-255.

Mankiw, N. G. and Miron, J. A. (1986) "The changing behavior of the term structure of interest rates", *Quarterly Journal of Economics*, **Vol. 101o. 2**, pp. 211-228.

Markowitz, H. M. (1959) *Portfolio Selection: Efficient Diversification of Investments*, John Wiley and Sons, New York.

Martin, J. D. and Klemkosky, R. C. (1975) "Evidence of heteroscedasticity in the market model", *Journal of Business*, **Vol. 48 No. 1**, pp. 81-86.

Miller, L. H. (1956) "Tables of Percentage Points of Kolmogorov Statistics", *Journal of the American Statistical Association*, **Vol. 51**, pp. 111-121.

Miller, M. H. and Scholes, M. (1972) "Rates of return in relation to risk: a re-examination of some recent findings", in Jensen, M. C., ed., *Studies in the Theory of Capital Markets*, Praeger Publishers, New York.

Miller, T. W. and Gessis, N. (1980) "Non stationarity and evaluation of mutual fund performance", *Journal of Financial and Quantitative Analysis*, **Vol. 15 no. 3**, pp. 639-654.

Moggridge, J. D., ed. (1973) *The General Theory and After*, New York, MacMillan.

Morgon, I. G. (1976) "Stock prices and heteroscedasticity", *Journal of Business*, Vol. 49 no. 4, pp. 496-508.

Page, E. S. 1954 "Continuous inspection schemes", *Biometrica*, Vol. 41, pp. 100-114.

Ohlson, J. and Rosenberg, B. (1982) "Systematic risk of the CRSP equal-weighted common stock index: a history estimated by stochastic-parameter regression", *Journal of Business*, Vol. 55 no. 1, pp. 121-145.

Pagan, A. R. (1980) "Some identification and estimation results for regression models with stochastically varying coefficients", *Journal of Econometrics*, Vol. 13, pp. 341-363.

Ploberger, W. (1989) "The local power of the cusum-sq test against heteroscedasticity" in Hackl, P., ed. *Statistical Analysis and Forecasting of Economic Structural Change*, Springer, Heidelberg, pp.127-133.

Ploberger, W., Krämer, W. and Kontrus, K. (1989) "A new test for structural stability in the linear regression model", *Journal of Econometrics*, Vol. 40, pp. 307-318.

Roenfeld, R., Griepentrog, G. L. and Pflaum, C. C., "Further evidence on the stationarity of beta coefficients", *Journal of Financial and Quantitative Analysis*, Vol. 13 no. 1, pp. 117-121.

Rosenberg, B. (1973) "Random coefficients models: the analysis of a cross section of time series by stochastically convergent parameter regression", *Annals of Social and Economic Measurement*, Vol. 2 no. 4, pp. 399-428.

Schaefer, S., Brealey, R., Hodges, S. and Thomas, H., "Alternative models of systematic risk" in Elton, E. and Gruber, M., eds, *International Capital Markets*, North Holland, Amsterdam, pp. 150-161.

Schneider, W., (1990) "Theory and application of Kalman filtering in economics and econometrics", Lecture notes from a 5 lecture cycle at the Stockholm School of Economics, March 1990.

Scott, E. and Brown, S. (1980) "Biased estimators and unstable betas", *Journal of Finance*, Vol. 35 no. 1, pp. 49-55.

Sharpe, W. F. (1964) "Capital asset prices: a theory of market equilibrium under conditions of risk", *Journal of Finance*, Vol. 19 no. 3, pp. 425-442.

Simonds, R. R., LaMotte, L. R. and McWhorter, A. Jr. (1986) "Testing for non-stationarity of market risk: an exact test and power considerations", *Journal of Financial and Quantitative Analysis*, Vol. 21 no. 2, pp. 209-220.

Stenuis, M. (1988) "Volatility and time–varying risk premiums in the stock market", University of Helsinki, Dept. of Economics, Discussion paper, no. 266.

Sunder, S. (1980) "Stationarity of market risk: random coefficients test for individual stocks", *Journal of Finance*, Vol. 35 no. 4, pp. 883-896.

Swamy, P. A. V. B., Conway, R. K., and LeBlanc, M. R. (1988) "The stochastic coefficients approach to econometric modeling. Part 1: A critique of fixed coefficient models", *Journal of Agricultural Economics Research*, Vol. 40 no. 2, pp. 2-10.

Szeto, M. W. (1973) "Estimation of the volatility of securities in the stock market by Kalman filtering techniques", in *Fourteenth Joint Automatic Control Conference of the American Automatic Control Council*, preprints of technical papers, Columbus Ohio, pp. 302-310.

Tesfatsion, L. and Veitch, J. M. (1990) "U.S. money demand instability:a flexible least squares approach", *Journal of Economic Dynamics and Control*, Vol. 14 no. 5, pp. 171-183.

Theobald, M. (1981) "Beta stationarity and estimation period: some analytical results", *Journal of Financial and Quantitative Analysis*, Vol. 16 no. 5, pp. 747-757.

Varian, H. R. (1993) *Intermediate Microeconomics*, W. W. Norton & Co., New York.

Vasicek, O. A. (1973) "A note on using cross-section information in Bayesian estimation of security betas", *Journal of Finance*, Vol. 28 no. 5, pp. 1233-1239.

Watson, M. W. and Engle, R. F. (1985) "Testing for regression coefficient stability with a stationary AR(1) alternative", *The Review of Economics and Statistics*, Vol. 67 no. 2, pp. 341-346.

Wells, C. (1978) *Optimal Fiscal and Monetary Policy*, CWK Gleerup, Lund.

Wells, C. (1990) "Beta coefficients on the Stockholm Exchange", *Beta*, **Vol. 4 no. 2**, pp. 45-61 (*in Swedish*).

Wells, C. (1992) "A test of the rational expectations and the permanent income hypothesis using Swedish data, 1963-87", in Velupillai, K., ed., *Nonlinearities, Disequilibria and Simulation*, MacMillan, London, pp. 163-196.

Wells, C., "Variable betas on the Stockholm Exchange", *Applied Financial Economics*, **Vol. 4 no. 1**, pp. 75-92.

Westlund, A. H. and Törnkvist, B. (1989) "On the identification of time for structural changes by MOSUM-SQ and CUSUM-SQ procedures", in Hackl, P., ed., *Statistical Analysis and Forecasting of Economic Structural Change*, Springer Verlag, pp. 97-126.

White, H. (1980) "A heteroscedastic-consistent covariance matrix and a direct test for heteroscedasticity", *Econometrica*, **Vol. 48 no. 4**, pp. 817-838.

White, H. (1982) "Maximum likelihood estimation of misspecified models", *Econometrica*, **Vol. 50 no. 1**, pp. 1-16.

Wolff, C. C. P. (1987) "Forward foreign exchange rates, expected stop rates, and premia: a signal-extraction approach", *Journal of Finance*, **Vol. 42 no. 2**, pp. 395-406.

Index

Advanced Studies in Theoretical and Applied Econometrics

1. J.H.P. Paelinck (ed.): *Qualitative and Quantitative Mathematical Economics.* 1982
ISBN 90-247-2623-9
2. J.P. Ancot (ed.): *Analysing the Structure of Econometric Models.* 1984
ISBN 90-247-2894-0
3. A.J. Hughes Hallet (ed.): *Applied Decision Analysis and Economic Behaviour.* 1984
ISBN 90-247-2968-8
4. J.K. Sengupta: *Information and Efficiency in Economic Decision.* 1985
ISBN 90-247-3072-4
5. P. Artus and O. Guvenen (eds.), in collaboration with F. Gagey: *International Macroeconomic Modelling for Policy Decisions.* 1986   ISBN 90-247-3201-8
6. M.J. Vilares: *Structural Change in Macroeconomic Models.* Theory and Estimation. 1986
ISBN 90-247-3277-8
7. C. Carraro and D. Sartore (eds.): *Development of Control Theory for Economic Analysis.* 1987
ISBN 90-247-3345-6
8. D.P. Broer: *Neoclassical Theory and Empirical Models of Aggregate Firm Behaviour.* 1987
ISBN 90-247-3412-6
9. A. Italianer: *Theory and Practice of International Trade Linkage Models.* 1986
ISBN 90-247-3407-X
10. D.A. Kendrick: *Feedback.* A New Framework for Macroeconomic Policy. 1988
ISBN 90-247-3593-9; Pb: 90-247-3650-1
11. J.K. Sengupta and G.K. Kadekodi (eds.): *Econometrics of Planning and Efficiency.* 1988
ISBN 90-247-3602-1
12. D.A. Griffith: *Advanced Spatial Statistics.* Special Topics in the Exploration of Quantitative Spatial Data Series. 1988   ISBN 90-247-3627-7
13. O. Guvenen (ed.): *International Commodity Market Models and Policy Analysis.* 1988
ISBN 90-247-3768-0
14. G. Arbia: *Spatial Data Configuration in Statistical Analysis of Regional Economic and Related Problems.* 1989   ISBN 0-7923-0284-2
15. B. Raj (ed.): *Advances in Econometrics and Modelling.* 1989 ISBN 0-7923-0299-0
16. A. Aznar Grasa: *Econometric Model Selection.* A New Approach. 1989
ISBN 0-7923-0321-0
17. L.R. Klein and J. Marquez (eds.): *Economics in Theory and Practice.* An Eclectic Approach. Essays in Honor of F. G. Adams. 1989   ISBN 0-7923-0410-1
18. D.A. Kendrick: *Models for Analyzing Comparative Advantage.* 1990
ISBN 0-7923-0528-0
19. P. Artus and Y. Barroux (eds.): *Monetary Policy.* A Theoretical and Econometric Approach. 1990   ISBN 0-7923-0626-0

Advanced Studies in Theoretical and Applied Econometrics

20. G. Duru and J.H.P. Paelinck (eds.): *Econometrics of Health Care.* 1990
ISBN 0-7923-0766-6

21. L. Phlips (ed.): *Commodity, Futures and Financial Markets.* 1991
ISBN 0-7923-1043-8

22. H.M. Amman, D.A. Belsley and L.F. Pau (eds.): *Computational Economics and Econometrics.* 1992 ISBN 0-7923-1287-2

23. B. Raj and J. Koerts (eds.): *Henri Theil's Contributions to Economics and Econometrics.* Vol. I: Econometric Theory and Methodology. 1992
ISBN 0-7923-1548-0

24. B. Raj and J. Koerts (eds.): *Henri Theil's Contributions to Economics and Econometrics.* Vol. II: Consumer Demand Analysis and Information Theory. 1992
ISBN 0-7923-1664-9

25. B. Raj and J. Koerts (eds.): *Henri Theil's Contributions to Economics and Econometrics.* Vol. III: Economic Policy and Forecasts, and Management Science. 1992 ISBN 0-7923-1665-7
Set (23-25) ISBN 0-7923-1666-5

26. P. Fisher: *Rational Expectations in Macroeconomic Models.* 1992
ISBN 0-7923-1903-6

27. L. Phlips and L.D. Taylor (eds.): *Aggregation, Consumption and Trade.* Essays in Honor of H.S. Houthakker. 1992. ISBN 0-7923-2001-8

28. L. Mátyás and P. Sevestre (eds.): *The Econometrics of Panel Data.* Handbook of Theory and Applications. 1992 ISBN 0-7923-2043-3

29. S. Selvanathan: *A System-Wide Analysis of International Consumption Patterns.* 1993 ISBN 0-7923-2344-0

30. H. Theil in association with D. Chen, K. Clements and C. Moss: *Studies in Global Econometrics.* 1996 ISBN 0-7923-3660-7

31. P.J. Kehoe and T.J. Kehoe (eds.): *Modeling North American Economic Integration.* 1995 ISBN 0-7923-3751-4

32. C. Wells: *The Kalman Filter in Finance.* 1996 ISBN 0-7923-3771-9

Kluwer Academic Publishers – Dordrecht / Boston / London

Printed in the United Kingdom
by Lightning Source UK Ltd.
104992UKS00001B/248